Math Kangaroo USA
Problems and Solutions

Grades 7 & 8

Even Years
2006–2024

Editor in Chief

Agata Gazal
Chief Editorial Officer for Math Kangaroo USA
Billings, MT

Reviewers and Contributors

Joanna Matthiesen
Chief Executive Officer for Math Kangaroo USA
Granger, IN

Izabela Szpiech
Chief Financial Officer for Math Kangaroo USA
Chicago, IL

Kasia Nalaskowska
Chief Information Officer for Math Kangaroo USA
Aurora, IL

Magdalena Teodorowicz
Chief Design Officer for Math Kangaroo USA
Cordova, TN

Professor Andrzej Zarach, Ph.D.
Math Content Reviewer
East Stroudsburg University, East Stroudsburg, PA

Book Design

Jossea K. Rilea
Designer at LX Design Lab
Saratoga Springs, NY

We would like to give special thanks to countless other people who contributed to the problems and solutions for this book since 1998. Primarily, a big thank you to the Math Kangaroo question writers from all over the world who are part of the AKSF organization (www.aksf.org). Math Kangaroo solution writers also include Math Kangaroo USA competition organizers and Math Kangaroo alumni. We would also like to thank the hundreds of educators who gave us feedback on the questions and solutions and finally the hundreds of thousands of students who took the Kangaroo challenge over the last two decades. Thank you all for your help in developing this book.

Copyright © 2025 by Math Kangaroo USA, NFP, Inc.
www.mathkangaroo.org

For additional copies of this book, please contact the publisher:
Math Kangaroo USA
info@mathkangaroo.org

ISBN: 9798989988365

Preface

Welcome to a world of math challenges and exciting problem-solving! Whether you're a student looking to sharpen your skills or a teacher eager to inspire young minds, this book is a treasure trove of stimulating questions and solutions.

Inside, you'll find 300 captivating problems, drawn from 10 years of the Math Kangaroo Competition (even years from 2006 to 2024), designed for 7th- and 8th-grade students. These questions are carefully selected at the annual Kangourou sans Frontières meeting, where mathematicians from over 100 countries gather to ensure each problem is engaging and age-appropriate. Each test consists of 30 questions, categorized by difficulty—easy, medium, and difficult—so that all students can find their challenge level.

This easy-to-use resource book is more than just a collection of problems. It's a journey into the world of math and logic, with visually appealing questions and insightful solutions that encourage children to think critically about the world around them. Problem-solving is a skill students use daily, often without realizing it. This book is designed to help them practice logical reasoning, enhance their math skills, and reflect on their problem-solving process.

We hope this book will inspire not only students who have a passion for math, but also educators who love to bring unconventional, thought-provoking challenges in the classroom. Whether you're seeking to improve mathematical thinking or simply enjoy the thrill of solving puzzles, we believe this book will provide both fun and valuable learning experiences.

Enjoy the journey!

Joanna Matthiesen, CEO

Color Key

Each test has 30 questions with 3 levels of difficulty

GREEN	YELLOW	RED
Easy	Medium	Difficult
Q 1-10	Q 11-20	Q 21-30
3 Points	4 Points	5 Points

TABLE OF CONTENTS

Part I PROBLEMS..7
2006..9
2008..15
2010..21
2012..27
2014..33
2016..39
2018..45
2020..51
2022..57
2024..63

Part II SOLUTIONS...69
2006..71
2008..77
2010..81
2012..87
2014..93
2016..99
2018..105
2020..113
2022..119
2024..127

Part III ANSWER KEY..135

Part I
Problems

2006

2006

3 Points Each

1 The Math Kangaroo contest has taken place in Europe every year since 1991. In year 2006 the _____ contest will take place.

(A) 15th
(B) 16th
(C) 17th
(D) 13th
(E) 14th

2 20 × (0 + 6) − (20 × 0) + 6 =

(A) 0
(B) 106
(C) 114
(D) 126
(E) 12

3 Point O is the center of the regular pentagon. What part of the whole pentagon is the shaded region?

(A) 10%
(B) 20%
(C) 25%
(D) 30%
(E) 40%

4 Two sides of a triangle are 120 and 130 inches long. Which of the following numbers could not be the length of the third side of the triangle in inches?

(A) 40
(B) 99
(C) 100
(D) 150
(E) 260

5 2006 students participated in a survey. The survey stated that 1500 of them participated in the Math Kangaroo contest, and 1200 of them participated in an English language contest. Out of the students who participated in the survey, how many participated in both contests if it is known that 6 people did not take part in either of the competitions?

(A) 300
(B) 500
(C) 600
(D) 700
(E) 1000

6 The solid shown in the picture is made out of two cubes, one with an edge 1 inch in length and the other with an edge 3 inches in length. What is the surface area of the solid in square inches?

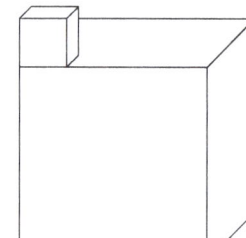

(A) 56
(B) 58
(C) 59
(D) 60
(E) 64

7 A bottle of a volume of $\frac{1}{2}$ liter is $\frac{3}{4}$ filled with juice. How much juice will be left in the bottle after pouring out $\frac{1}{5}$ of a liter?

(A) $\frac{1}{20}$ liter
(B) $\frac{3}{40}$ liter
(C) 0.13 liter
(D) $\frac{1}{8}$ liter
(E) The bottle will be empty.

8. Out of all possible isosceles triangles with sides of 7 and a base with a length expressed by a whole number, the triangle with the greatest perimeter was selected. This perimeter is equal to:

(A) 14
(B) 15
(C) 21
(D) 27
(E) 28

9. Granny baked dumplings for her grandchildren. If she gives each of her grandchildren 2 dumplings, she will still have 3 more dumplings left. However, if she gives each of them 3 dumplings, she will be 2 dumplings short. How many grandchildren does she have?

(A) 2
(B) 3
(C) 4
(D) 5
(E) 6

10. Out of which of the figures can you make the box shown in the picture?

(A)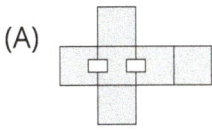

(B)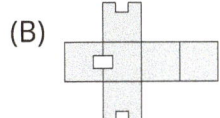

(C)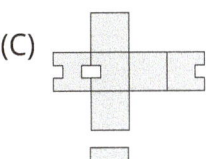

(D)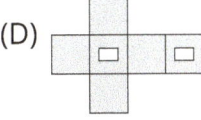

(E)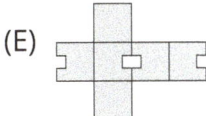

4 Points Each

11. How many whole numbers smaller than 100 can you get as a sum of nine consecutive integers?

(A) 13
(B) 12
(C) 11
(D) 10
(E) 9

12. In a certain month, three Tuesdays turned out to be on even days of the month. Which day of the week will be the 21st day of that month?

(A) Sunday
(B) Saturday
(C) Friday
(D) Thursday
(E) Wednesday

13. Mark, Matthew, and Peter were saving money to buy a tent. Mark had 60% of the needed sum, and Matthew gave 40% of the remaining amount. Peter gave 30 dollars to complete the sum. How much did the tent cost?

(A) $50
(B) $60
(C) $125
(D) $150
(E) $200

14 A group of aliens was traveling in a spaceship. Each one of them was dressed in a jumpsuit that was one of the following colors: green, orange, or blue. Each alien dressed in a green jumpsuit had two tentacles, each alien dressed in an orange suit had three tentacles, and each alien dressed in a blue jumpsuit had five tentacles. There were as many aliens dressed in green suits as those dressed in orange, and there were 10 more aliens dressed in blue than those dressed in green. All the aliens together had 250 tentacles. How many aliens were dressed in blue?

(A) 15
(B) 20
(C) 25
(D) 30
(E) 4

15 When Jumpy the Kangaroo jumps on his left foot, his jump is 2 feet long. When he jumps on his right foot, his jump is 4 feet long. If Jumpy jumps using both his feet, then his jump is 7 feet long. What is the smallest number of jumps that Jumpy must make to travel exactly 1000 feet?

(A) 140
(B) 144
(C) 175
(D) 176
(E) 150

16 The rectangle in the picture is divided into 7 squares. The sides of the gray squares are each 8 units long. What is the length of the side of the big white square?

(A) 16
(B) 18
(C) 20
(D) 24
(E) 30

17 A square of a positive number is 500% greater than that number. What number is it?

(A) 5
(B) 6
(C) 7
(D) 8
(E) 10

18 In isosceles triangle ABC with side AC = BC, angle BAC has been bisected by AD (see the picture), and AD = AB. What is the measure of angle ACB?

(A) 22°
(B) 30°
(C) 36°
(D) 45°
(E) 60°

19 Barbara is creating different squares using sticks of equal length. She labels them with numbers 1, 2, 3, etc. How many more sticks will she use to create 31st square than the 30th square?

(A) 124
(B) 148
(C) 61
(D) 254
(E) 120

20 The ones digit of a certain three-digit number is 2. If we move this digit to the beginning of that number, as a result we will get a three-digit number that is smaller than the original number by 36. What is the sum of the digits of this number?

(A) 1
(B) 10
(C) 7
(D) 9
(E) 5

5 Points Each

21 Helen drew a 5 × 5 square and marked the center of each small square. Afterwards, she drew obstacles and then she tested in how many ways it was possible to move from A to B in the shortest possible way while avoiding the obstacles and moving vertically or horizontally from center to center of each small square. How many such paths with the shortest length are there?

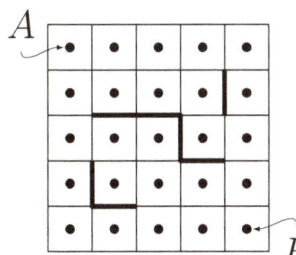

(A) 6
(B) 8
(C) 9
(D) 11
(E) 12

22 A train consists of an engine and five cars marked I, II, III, IV, and V. In how many ways can you rearrange the cars in such way that car I is always closer to the engine than car II?

(A) 120
(B) 60
(C) 48
(D) 30
(E) 10

23 What is the first digit of the smallest natural number with the sum of its digits equal to 2006?

(A) 1
(B) 3
(C) 5
(D) 6
(E) 8

24 How many isosceles triangles with an area of 1 have a side of length equal to 2?

(A) 0
(B) 1
(C) 2
(D) 3
(E) 4

25 Peter rides a bicycle from town P to town Q at a constant speed. If he increases his speed by 3 m/s, he will arrive at town Q 3 times faster. How many times faster will Peter arrive at town Q if he increases his speed by 6 m/s?

(A) 4
(B) 5
(C) 6
(D) 4.5
(E) 8

26 If the product of two integers equals $2^5 \times 3 \times 5^2 \times 7^3$, then their sum:

(A) can be divisible by 8.
(B) can be divisible by 3.
(C) can be divisible by 5.
(D) can be divisible by 49.
(E) cannot be divisible by any of the numbers 8, 3, 5, or 49.

27 Let ABCD be a square with a side of length equal to 12 inches. Points P, Q, R are the midpoints of sides BC, CD, and DA, respectively (see the figure). The area of the shaded region in square inches is:

(A) 96
(B) 72
(C) 60
(D) 54
(E) 48

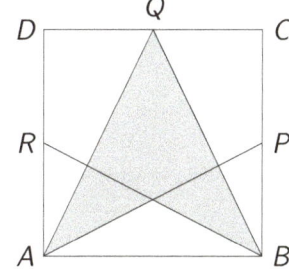

28. The first table shows 11 cards, each with 2 letters. The second table shows rearrangement of the cards from the first table. Which of the following could appear on the bottom row of the second table?

M	I	S	S	I	S	S	I	P	P	I
K	I	L	I	M	A	N	J	A	R	O

P	S	I	S	I	M	I	S	S	P	I

(A) ANJAMKILIOR
(B) RLIIMKOJNAA
(C) JANAMKILIRO
(D) RAONJMILIKA
(E) ANMAIKOLIRJ

---

29. What is the value of this expression:

$(1^2 + 2^2 + 3^2 + \ldots + 2005^2) -$
$(1 \times 3 + 2 \times 4 + 3 \times 5 + \ldots + 2004 \times 2006)$

(A) 2000
(B) 2004
(C) 2005
(D) 2006
(E) 0

---

30. Let x, y, and z be positive real numbers and let $x \geq y \geq z$ so that $x + y + z = 20.1$. Which of the following statements is always true?

(A) $xy < 99$
(B) $xy > 1$
(C) $xy \neq 75$
(D) $xy \neq 25$
(E) None of the statements is true.

2008

2008

3 Points Each

1 Which of the numbers is the largest?

(A) (1 × 2) × (2007 × 2008)
(B) (1 + 2) × (2007 × 2008)
(C) (1 × 2) × (2007 + 2008)
(D) (1 + 2) × (2007 + 2008)
(E) (1 + 2) + (2007 + 2008)

2 There are 9 boys and 13 girls in one class. Half of the children in the class have a cold. At the very least, how many of the girls have a cold?

(A) 0
(B) 1
(C) 2
(D) 3
(E) 4

3 If
$$\frac{1}{1+\frac{1}{x}} = 2$$
then the result of
$$\frac{1}{1+\frac{1}{1+\frac{1}{x}}}$$
is equal to:

(A) $\frac{3}{2}$
(B) $\frac{1}{3}$
(C) $\frac{2}{3}$
(D) 4
(E) $\frac{1}{2}$

4 The blades on the front of a windmill rotate at a constant speed. The whole set makes a full rotation in 50 seconds. How many blades are there, if a sensor mounted on the top of the windmill notes that a blade passes every 10 seconds?

(A) 2
(B) 3
(C) 5
(D) 10
(E) 50

5 The numbers 2, 3, 4, and one more number are written in the cells of a 2 × 2 table. It is known that the sum of the numbers in the first row is equal to 9, and the sum of the numbers in the second row is equal to 6. The unknown number is:

(A) 4
(B) 5
(C) 6
(D) 7
(E) 8

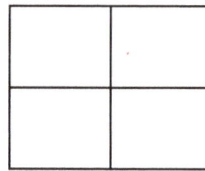

6 The sum of digits of the sum of the digits of 2008 is equal to:

(A) 2
(B) 6
(C) 8
(D) 10
(E) 1

7 The triangle and the square have the same perimeter. What is the perimeter of the whole figure (a pentagon)?

(A) 12 cm
(B) 24 cm
(C) 28 cm
(D) 32 cm
(E) 20 cm

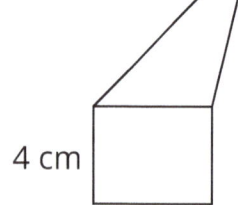

8. The florist had 102 roses: 24 white, 42 red, and 36 yellow. At most, how many identical bunches can she make if she wants to use all of the flowers?

(A) 4
(B) 6
(C) 8
(D) 10
(E) 12

9. A cube has all its corners cut off, as shown in the picture. How many sides does the resulting solid have?

(A) 10
(B) 18
(C) 12
(D) 16
(E) 14

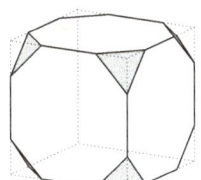

10. Three lines intersect in one point. Two angles are given. What is the measure of the shaded angle?

(A) 52°
(B) 53°
(C) 54°
(D) 55°
(E) 56°

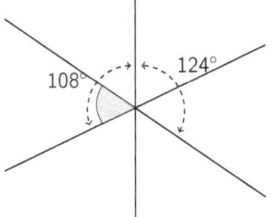

4 Points Each

11. How many squares can be drawn by joining the dots with line segments?

(A) 2
(B) 3
(C) 4
(D) 5
(E) 6

12. Dan has 9 coins, each worth 2 cents. His sister Ann has 8 coins, each worth 5 cents. What is the least number of coins they should exchange with each other in order for each of them to have the same amount of money?

(A) 4
(B) 5
(C) 8
(D) 12
(E) It is impossible to do.

13. In the year 2008, the ones digit is equal to four times the thousands digit. What is the minimum number of years that has to pass for this situation to occur again?

(A) 10
(B) 20
(C) 100
(D) 2008
(E) different answer

14. How many pairs of digits a, b from the set {0, 1, 2, 3, 4, 5, 6, 7, 8, 9} make the following equation true: $a \times b = 10 + a$?

(A) 0
(B) 1
(C) 2
(D) 3
(E) 4

15. The British mathematician August de Morgan claimed that he was x years old in the year of x^2. He is known to have died in 1899. When was he born?

(A) 1806
(B) 1848
(C) 1849
(D) 1899
(E) different answer

16 Triangle ABC is isosceles, AB = AC, ∠BPC = 120°, and ∠ABP = 50°. What is the measure of angle PBC?

(A) 5°
(B) 10°
(C) 15°
(D) 20°
(E) 25°

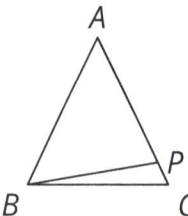

17 Two identical rectangular pieces of paper were cut into two pieces. The first one resulted in two rectangles of perimeters of 40 cm each. The second one resulted in two rectangles with perimeters of 50 cm each. What is the perimeter of each original rectangle?

(A) 40 cm
(B) 50 cm
(C) 60 cm
(D) 80 cm
(E) 90 cm

18 Points A, B, C, and D are marked on a straight line in a certain order. It is known that AB = 13, BC = 11, CD = 14, and DA = 12. What is the distance between the two points which are farthest from each other?

(A) 25
(B) 14
(C) 38
(D) 50
(E) 39

19 Four tangent congruent circles of radius 6 cm are inscribed in a rectangle. If P is a vertex and Q and R are points of tangency, what is the area of triangle PQR?

(A) 27 cm²
(B) 45 cm²
(C) 54 cm²
(D) 108 cm²
(E) 180 cm²

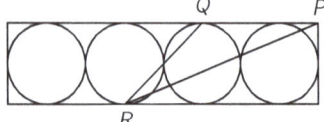

20 One of the faces of a cube is cut along its diagonals (see the figure). Which of the following diagrams cannot be a representation of this cube after it is unfolded?

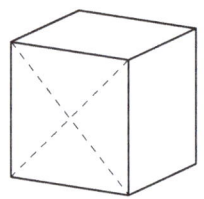

1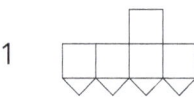

2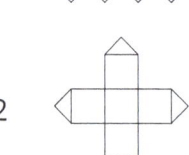

3

4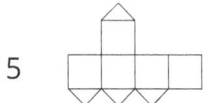

5

(A) 1 and 3
(B) 1 and 5
(C) 3 and 4
(D) 3 and 5
(E) 2 and 4

5 Points Each

21 A wooden cube with the dimension 5 × 5 × 5 has been constructed by gluing together 5³ unit cubes. Cleo took a picture of the cube in such a way that the largest possible number of unit cubes was visible in the picture. How many unit cubes were visible in the picture?

(A) 75
(B) 74
(C) 60
(D) 61
(E) 62

22. Kangaroo came up with a new operation ∗ in the set of positive natural numbers. He gave a few examples:

$2 * 3 = (2 + 1) \times 3 = 9$;
$4 * 2 = (4 + 3 + 2 + 1) \times 2 = 20$;
$3 * 5 = (3 + 2 + 1) \times 5 = 30$.

What is the answer to 6 ∗ 5?

(A) 30
(B) 90
(C) 105
(D) 210
(E) 315

23. In an isosceles triangle ABC, the bisector CD of the angle C is equal to the base BC. What is the measure of angle CDA?

(A) 90°
(B) 100°
(C) 108°
(D) 120°
(E) It cannot be to determined.

24. In the equation KAN − GAR = OO, each letter represents a certain digit (different letters represent different digits, the same letters represents the same digits). Find the largest possible value of the number KAN.

(A) 987
(B) 876
(C) 865
(D) 864
(E) 785

25. In a certain class, the girls make up more than 45% but less than 50% of the whole class. What is the smallest possible number of girls in that class?

(A) 3
(B) 4
(C) 5
(D) 6
(E) 7

26. A certain boy always tells the truth on Thursdays and Fridays, always lies on Tuesdays, and randomly tells the truth or lies on other days of the week. On seven consecutive days he was asked what his name was, and on the first six days he gave the following answers in this order: John, Bob, John, Bob, Peter, Bob. What did he answer on the seventh day?

(A) John
(B) Bob
(C) Peter
(D) Kate
(E) different answer

27. A truck, traveling at a constant speed, traveled from town A to town B in 1 hour and 30 minutes, and then from town B to town C in 1 hour. Following the same route, a car, also traveling at a constant speed, traveled from town A to town B in 1 hour. How long did it take the car to travel from town B to town C?

(A) 45 minutes
(B) 40 minutes
(C) 35 minutes
(D) 30 minutes
(E) 90 minutes

28. Given are two sets of four-digit numbers: set A, which consists of the numbers the product of whose digits is equal to 25, and set B, which consists of the numbers the product of whose digits is equal to 15. Which set consists of more numbers, and many times more numbers are there?

(A) Set A has 5/3 times as many elements as set B.
(B) Set A has 2 times as many elements as set B.
(C) Set B has 5/3 times as many elements as set A.
(D) Set B has 2 times as many elements as set A.
(E) The numbers of elements in both sets are equal.

29. Triangles ABC and ABD are inscribed in a circle, shown on the figure. We know that ∠BAC = 45° and ∠BAD = 135°. Which of the following statements is true about segments BC and BD?

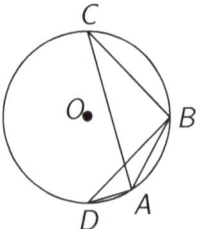

(A) BC > BD
(B) BC < BD
(C) BC = BD
(D) BC = 2BD
(E) BD = 2BC

30. There are seven cards in a box. Numbers from 1 to 7 are written on these cards, exactly one number on each card. The first sage takes 3 cards at random from the box and the second sage takes 2 cards. 2 cards are left in the box. Then, the first sage says to the second one: "I know that the sum of the numbers of your cards is even." The sum of the numbers on the cards of the first sage is equal to:

(A) 10
(B) 12
(C) 6
(D) 9
(E) 15

2010

2010

3 Points Each

1 The sum 12 + 23 + 34 + 45 + 56 + 67 + 78 + 89 equals

(A) 389
(B) 396
(C) 404
(D) 405
(E) 504

2 How many axes of symmetry does the figure have?

(A) 0
(B) 1
(C) 2
(D) 4
(E) infinitely many

3 A shipment of toys is being prepared. Each toy is placed into a cube box. Exactly eight such boxes fit tightly into a larger cube carton. How many boxes with toys are on the bottom layer of the large carton?

(A) 1
(B) 2
(C) 3
(D) 4
(E) 5

4 Each pair of adjacent sides of the figure shown in the picture is perpendicular. What is the perimeter of this figure?

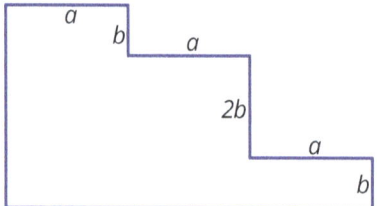

(A) $6a + 8b$
(B) $3a + 4b$
(C) $3a + 8b$
(D) $6a + 4b$
(E) $6a + 6b$

5 The six points shown in the picture are the vertices of a regular hexagon. A geometric figure that cannot be obtained by connecting some of these points with line segments is a

(A) trapezoid
(B) square
(C) kite
(D) right triangle
(E) obtuse triangle

6 The arithmetic mean of nine consecutive whole numbers equals 2006. The largest of these nine numbers is

(A) 2007
(B) 2009
(C) 2010
(D) 2011
(E) 2012

7 Grandmother baked cookies for her grandchildren, some of whom were going to visit her that day. She wanted each of her grandchildren to have the same number of cookies, but she forgot if three, five, or six grandchildren were coming. So, she baked enough cookies so that there would be an equal number for every grandchild regardless of how many visited. Which of these could be the number of cookies grandmother baked?

(A) 12
(B) 15
(C) 18
(D) 24
(E) 30

8 On the bottom of a rectangular box with a 5 cm × 5 cm square base are 7 blocks measuring 3 cm × 1 cm each. What is the smallest number of blocks that must be moved to fit an eighth block of the same dimensions on the bottom of the box?

(A) 2
(B) 3
(C) 4
(D) 5
(E) This is impossible.

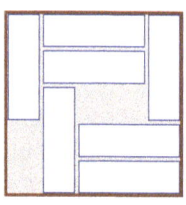

9. Jenny is drawing 2 × 2 squares. She divides each square into four square tiles and paints each of these tiles either blue or green. Squares which can be roatated and placed on top of one another so that the tiles match exactly are considered identical. (The figure shows identical squares.)

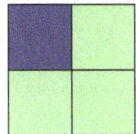 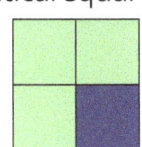 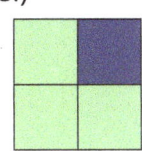 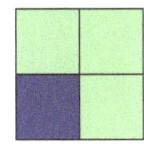

The largest number of nonidentical squares possible is

(A) 5
(B) 6
(C) 7
(D) 8
(E) 9

10. The difference between the sum of the first one hundred consecutive positive even numbers and the sum of the first one hundred consecutive positive odd numbers equals

(A) 0
(B) 50
(C) 100
(D) 10100
(E) 15150

4 Points Each

11. A lumberjack chopped wood, each time cutting a piece of wood in half. When he finished he observed that after making 53 cuts he had 72 pieces of wood. With how many pieces did he begin?

(A) 17
(B) 18
(C) 19
(D) 20
(E) 21

12. The smallest two-digit number that is not the sum of three different one-digit numbers is

(A) 10
(B) 15
(C) 23
(D) 25
(E) 28

13. A blacksmith needs 18 minutes to join three smaller chain segments into one larger chain segment. How much time does the blacksmith need to join six smaller chain segments if it takes him the same amount of time to join two chains of any length?

(A) 27 min
(B) 30 min
(C) 36 min
(D) 45 min
(E) 54 min

14. Some angle measures of the quadrilateral ABCD are shown in the picture. In addition, AD = BC. Find the measure of angle ABC.

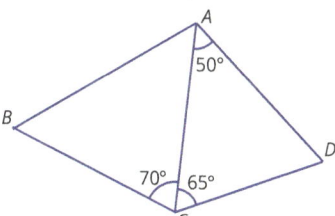

(A) 50°
(B) 55°
(C) 60°
(D) 65°
(E) It is impossible to compute.

15. A large box contains 50 blocks. Some are white, some are red, and some are blue. There are 11 times as many white blocks as there are blue blocks. There are fewer red blocks than there are white blocks, but there are more red blocks than there are blue blocks. By how much is the number of red blocks smaller than the number of white blocks?

(A) 14
(B) 11
(C) 19
(D) 22
(E) 30

16 Andrew wrapped a wire around a notched board. The picture shows the front side of the board. Which of the pictures shows the back side of the board?

(A)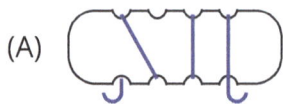

(B)

(C)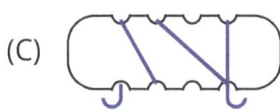

(D)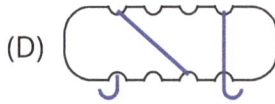

(E)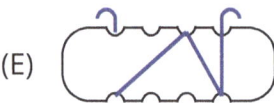

17 In the picture *PQRS* is a square. The area of the shaded region is equal to half the area of rectangle *ABCD*. What is the length of segment *PX*?

(A) 1
(B) 1.5
(C) 2
(D) 2.5
(E) 4

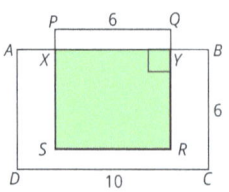

18 The smallest number of straight lines that will divide a plane into exactly 5 regions is

(A) 3
(B) 4
(C) 5
(D) 6
(E) 2010

19 If $a - 1 = b + 2 = c - 3 = d + 4 = e - 5$, then the largest number among *a, b, c, d*, and *e* is

(A) *a*
(B) *b*
(C) *c*
(D) *d*
(E) *e*

20 The figure is composed of arcs of semicircles with radii of 2 cm, 4 cm, and 8 cm. What fraction of the figure is shaded?

(A) $\frac{1}{3}$
(B) $\frac{1}{4}$
(C) $\frac{1}{5}$
(D) $\frac{3}{4}$
(E) $\frac{2}{3}$

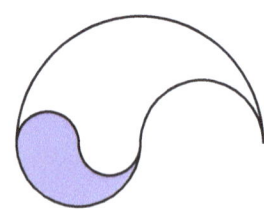

5 Points Each

21 Which of the following numbers cannot be equal to the sum of at least two positive consecutive whole numbers?

(A) 14
(B) 24
(C) 64
(D) 103
(E) 2010

22 A barter market trades poultry according to the rates shown in the table. What is the minimum number of hens necessary to trade for one turkey, one goose, and one chicken?

Rates of Exchange		
1 turkey	=	5 chickens
1 goose + 2 hens	=	3 chickens
4 hens	=	1 goose

(A) 18
(B) 17
(C) 16
(D) 15
(E) 14

23 Sides AB and AC of the triangle ABC are the diameters of two circles. These circles intersect at points A and P. Then, point P is necessarily

(A) the center of a circle inscribed in triangle ABC.
(B) the midpoint of side BC.
(C) inside triangle ABC.
(D) outside triangle ABC.
(E) the point of intersection of the height of triangle ABC at angle A with side BC.

---

24 Each of 18 cards is numbered with either a 4 or a 5. It turns out that the sum of all the numbers is divisible by 17. How many cards are labeled with a 4?

(A) 4
(B) 5
(C) 6
(D) 7
(E) 9

---

25 Two tangent circles with radii of 9 cm and 17 cm are placed inside a rectangle as shown. Find the length of the rectangle given that its width is 50 cm.

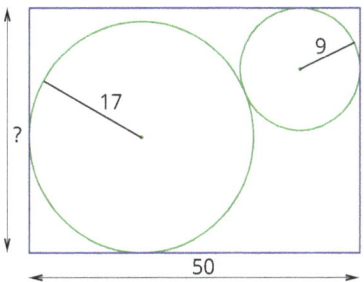

(A) $20\sqrt{2}$ cm
(B) 30 cm
(C) 36 cm
(D) $10\sqrt{13}$ cm
(E) 40 cm

---

26 A certain town is inhabited only by knights and liars. Knights always tell the truth and liars always lie. One day, several of the townspeople met in a room and three of them made the following statements:

- First person: "There are no more than three of us in the room. Each of us is a liar."
- Second person: "There are at most four people in the room. Not every one of us is a liar."
- Third person: "There are five people in the room. Three of us are liars."

How many people were in the room and how many of them were liars?

(A) 3 people, 1 liar
(B) 4 people, 1 liar
(C) 4 people, 2 liars
(D) 5 people, 2 liars
(E) 5 people, 3 liars

---

27 A kangaroo has a large collection of small 1 × 1 × 1 cubes. Each cube is painted one color. The kangaroo wants to construct a cube using 27 small cubes in such a way that any two small cubes sharing a vertex are of different colors. At least how many different colors must the kangaroo use?

(A) 7
(B) 8
(C) 9
(D) 12
(E) 27

---

28 The large triangle shown in the picture was divided into 36 small equilateral triangles, each with an area of 1. Triangle ABC has an area of

(A) 11
(B) 12
(C) 15
(D) 9
(E) 10

29 If the least common multiple of the numbers 24 and x is smaller than the least common multiple of the numbers 24 and y, then $\frac{y}{x}$ cannot equal

(A) $\frac{7}{8}$
(B) $\frac{8}{7}$
(C) $\frac{2}{3}$
(D) $\frac{6}{7}$
(E) $\frac{7}{6}$

---

30 Each of the five students whose birthday is today brought candy to class. It turned out that each had a different number of pieces of candy and that any three of these students had more candy than the other two. What is the smallest possible number of pieces of candy that they had altogether?

(A) 20
(B) 25
(C) 30
(D) 35
(E) 40

2012

2012

3 Points Each

1. Four chocolate bars cost 6 dollars more than one chocolate bar. What is the cost of one chocolate bar?

   (A) 1 dollar
   (B) 2 dollars
   (C) 3 dollars
   (D) 4 dollars
   (E) 5 dollars

2. 11.11 − 1.111 =

   (A) 9.009
   (B) 9.0909
   (C) 9.99
   (D) 9.999
   (E) 10

3. A watch is placed face up on a table so that its minute hand points northeast. How many minutes pass before the minute hand points northwest for the first time?

   (A) 45
   (B) 40
   (C) 30
   (D) 20
   (E) 15

4. Mary has a pair of scissors and five cardboard letters. She cuts each letter exactly once (along a straight line) so that it falls apart into as many pieces as possible. Which letter falls apart into the most pieces?

   (A)
   (B)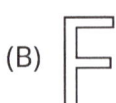
   (C)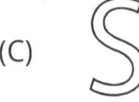
   (D)
   (E)

5. A dragon has five heads. Every time a head is chopped off, five new heads grow. If six heads are chopped off one by one, how many heads will the dragon have at the end?

   (A) 25
   (B) 28
   (C) 29
   (D) 30
   (E) 35

6. In which of the following expressions can we replace each occurrence of the number 8 by the same positive number (other than 8) and obtain the same result?

   (A) $(8 + 8) \div 8 + 8$
   (B) $8 \times (8 + 8) \div 8$
   (C) $8 + 8 - 8 + 8$
   (D) $(8 + 8 - 8) \times 8$
   (E) $(8 + 8 - 8) \div 8$

7. Each of the nine paths in a park is 100 m long. Ann wants to go from A to B without going along any path more than once. What is the length of the longest route she can choose?

   (A) 900 m
   (B) 800 m
   (C) 700 m
   (D) 600 m
   (E) 400 m

   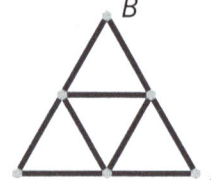

8. The diagram shows two triangles. In how many ways can you choose two vertices, one on each triangle, so that the straight line through the vertices does not cross either triangle?

   (A) 1
   (B) 2
   (C) 3
   (D) 4
   (E) more than 4

9. Werner folds a sheet of paper as shown in the figure and then makes two straight cuts with a pair of scissors. He then opens up the paper again. Which of the following shapes cannot be the result?

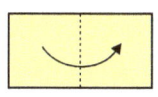 

(A)

(B)

(C)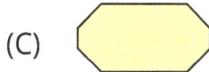

(D)

(E)

10. A rectangular prism is made up of four pieces, as shown. Each piece consists of four cubes and is a single color. What is the shape of the white piece?

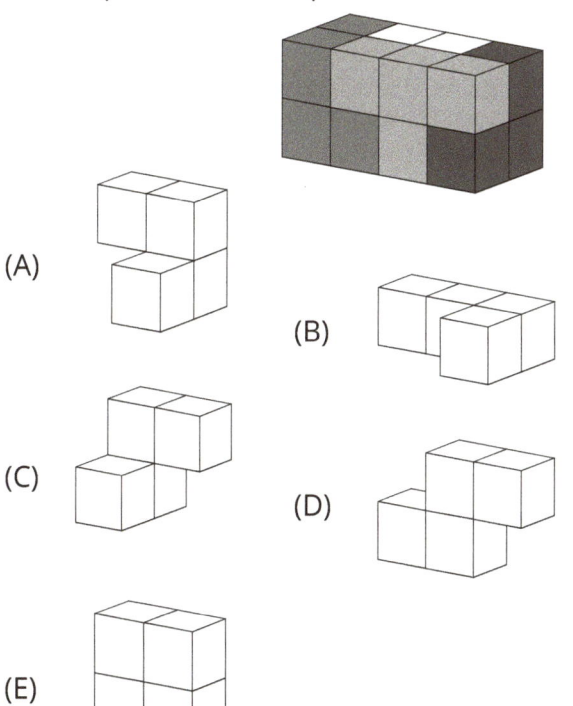

4 Points Each

11. Kanga forms two 4-digit natural numbers using each of the digits 1, 2, 3, 4, 5, 6, 7, and 8 exactly once. Kanga wants the sum of the two numbers to be as small as possible. What is the value of this smallest possible sum?

(A) 2468
(B) 3333
(C) 3825
(D) 4734
(E) 6912

12. Mrs. Gardner grows peas and strawberries. This year she has changed the rectangular pea bed to a square by lengthening one of its sides by 3 meters. As a result of this change, the area of the strawberry bed was reduced by 15 m². What was the area of the pea bed before the change?

(A) 5 m²
(B) 9 m²
(C) 10 m²
(D) 15 m²
(E) 18 m²

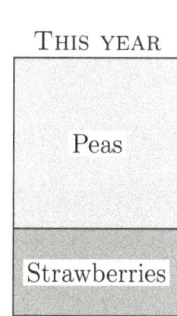

13. Barbara wants to complete the diagram by inserting three numbers, one in each empty cell. She wants the sum of the first three numbers to be 100, the sum of the three middle numbers to be 200 and the sum of the last three numbers to be 300. What number should Barbara insert in the middle cell of the diagram?

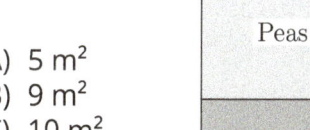

(A) 50
(B) 60
(C) 70
(D) 75
(E) 100

14 In the figure, what is the value of *x*?

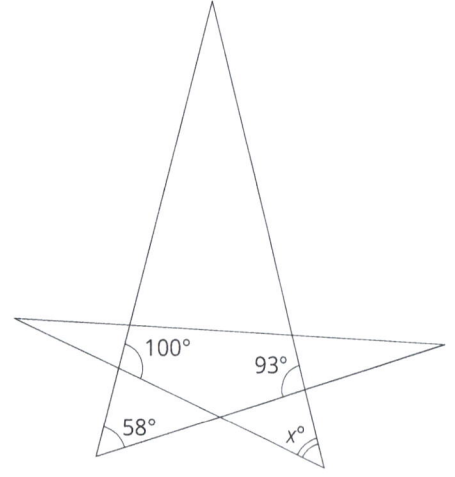

(A) 35
(B) 42
(C) 51
(D) 65
(E) 109

---

15 Four cards each have a number written on one side and a phrase written on the other. The four phrases are "divisible by 7," "prime," "odd," and "greater than 100," and the four numbers are 2, 5, 7, and 12. On each card, the number does not correspond to the phrase on the other side. What number is written on the same card as the phrase "greater than 100?"

(A) 2
(B) 5
(C) 7
(D) 12
(E) It is impossible to determine.

---

16 Three small equilateral triangles of the same size are cut from the corners of a larger equilateral triangle with sides of 6 cm, as shown. The sum of the perimeters of the three small triangles is equal to the perimeter of the remaining gray hexagon. What is the side length of the small triangles?

(A) 1 cm
(B) 1.2 cm
(C) 1.25 cm
(D) 1.5 cm
(E) 2 cm

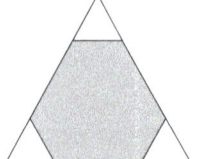

---

17 A piece of cheese was cut into a large number of pieces. During the course of the day, a number of mice came and stole some pieces, watched by the lazy cat Ginger. Ginger noticed that each mouse stole a different number of pieces, each of which was less than 10, and that no mouse stole exactly twice as many pieces as any other mouse. What is the largest number of mice that Ginger could have seen stealing cheese?

(A) 4
(B) 5
(C) 6
(D) 7
(E) 8

---

18 At the airport there is a moving walkway 500 meters long, which moves at a speed of 4 km/hour. Ann and Bill step on the walkway at the same time. Ann walks at a speed of 6 km/hour on the walkway while Bill stands still. When Ann comes to the end of the walkway, how far ahead of Bill is she?

(A) 100 m
(B) 160 m
(C) 200 m
(D) 250 m
(E) 300 m

---

19 A magical talking square originally has sides of length 8 cm. If he tells the truth, then his sides become 2 cm shorter. If he lies, then his perimeter doubles. He makes four statements, two true and two false, in some order. What is the largest possible perimeter of the square after the four statements?

(A) 28
(B) 80
(C) 88
(D) 112
(E) 120

20. A cube is rolled on a plane so that it turns around its edges. It begins at position 1, and is rolled so that one of its faces touches the plane in positions 2, 3, 4, 5, 6, and 7, in that order, as shown. Which two of these positions were occupied by the same face of the cube?

(A) 1 and 7
(B) 1 and 6
(C) 1 and 5
(D) 2 and 7
(E) 2 and 6

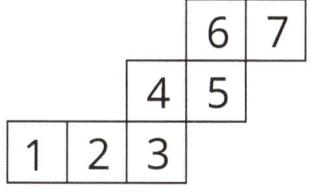

5 Points Each

21. Rick has five cubes. When he arranges them from smallest to largest, the difference between the heights of any two neighboring cubes is 2 cm. The largest cube is as high as a tower built from the two smallest cubes. How high is a tower built from all five cubes?

(A) 6 cm
(B) 14 cm
(C) 22 cm
(D) 44 cm
(E) 50 cm

22. In the diagram, ABCD is a square, M is the midpoint of AD, and MN is perpendicular to AC. What is the ratio of the area of the shaded triangle MNC to the area of the square?

(A) 1:6
(B) 1:5
(C) 7:36
(D) 3:16
(E) 7:40

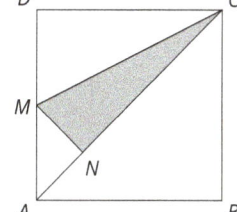

23. The tango is danced in pairs, each consisting of one man and one woman. At a dance evening no more than 50 people are present. At one moment $\frac{3}{4}$ of the men are dancing with $\frac{4}{5}$ of the women. How many people are dancing at that moment?

(A) 20
(B) 24
(C) 30
(D) 32
(E) 46

24. David wants to arrange the twelve numbers from 1 to 12 in a circle so that any two neighboring numbers differ by either 2 or 3. Which of the following pairs of numbers have to be neighbors?

(A) 5 and 8
(B) 3 and 5
(C) 7 and 9
(D) 6 and 8
(E) 4 and 6

25. Some three-digit integers have the following property: if you remove the first digit of the number, you get a perfect square; if instead you remove the last digit of the number, you also get a perfect square. What is the sum of all the three-digit integers with this curious property?

(A) 1013
(B) 1177
(C) 1465
(D) 1993
(E) 2016

26 A book contains 30 stories, each starting on a new page. The lengths of the stories are 1, 2, 3, ..., 30 pages. The first story starts on the first page. What is the largest number of stories that can start on an odd-numbered page?

(A) 15
(B) 18
(C) 20
(D) 21
(E) 23

27 An equilateral triangle starts in a given position and is rotated into new positions in a sequence of steps. At each step it is rotated about its center, first by 3°, then by a further 9°, then by a further 27°, and so on (at the *n*-th step it is rotated by a further $(3^n)°$). How many different positions, including the initial position, will the triangle occupy? (Two positions are considered equal if the triangle covers the same part of the plane.)

(A) 3
(B) 4
(C) 5
(D) 6
(E) 360

28 A rope is folded in half, then in half again, and then in half again. Finally, the folded rope is cut through, forming several strands. The lengths of two of the strands are 4 m and 9 m. Which of the following could not have been the length of the whole rope?

(A) 52 m
(B) 68 m
(C) 72 m
(D) 88 m
(E) All the previous are possible.

29 A triangle is divided into four triangles and three quadrilaterals by three straight line segments (see the figure). The sum of the perimeters of the three quadrilaterals is equal to 25 cm. The sum of the perimeters of the four triangles is equal to 20 cm. The perimeter of the whole triangle is equal to 19 cm. What is the sum of the lengths of the three straight line segments in cm?

(A) 11
(B) 12
(C) 13
(D) 15
(E) 16

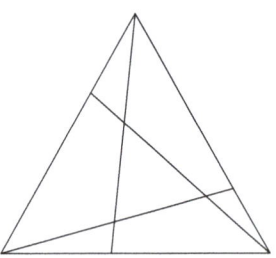

30 A positive number needs to be placed in each cell of the 3 × 3 grid shown, so that in each row and each column the product of the three numbers is equal to 1, and in each 2 × 2 square the product of the four numbers is equal to 2. What number should be placed in the central cell?

(A) 16
(B) 8
(C) 4
(D) $\frac{1}{4}$
(E) $\frac{1}{8}$

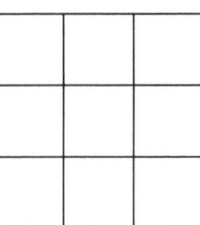

2014

2014

3 Points Each

1. Each year, the date of the Kangaroo competition is the third Thursday of March. What is the latest possible date of the competition in any year?

   (A) March 14th
   (B) March 15th
   (C) March 20th
   (D) March 21st
   (E) March 22nd

2. How many quadrilaterals of any size are shown in the figure?

   (A) 0
   (B) 1
   (C) 2
   (D) 4
   (E) 5

   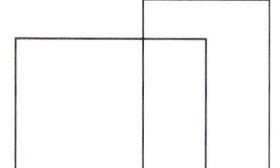

3. What is the result of 2014 × 2014 ÷ 2014 − 2014?

   (A) 0
   (B) 1
   (C) 2013
   (D) 2014
   (E) 4028

4. The area of rectangle ABCD is 10. Points M and N are the midpoints of sides AD and BC. What is the area of quadrilateral MBND?

   (A) 0.5
   (B) 5
   (C) 2.5
   (D) 7.5
   (E) 10

   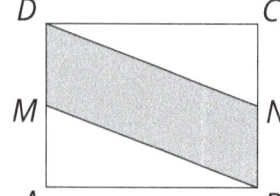

5. The product of two numbers is 36 and their sum is 37. What is their difference?

   (A) 1
   (B) 4
   (C) 10
   (D) 26
   (E) 35

6. Wanda has several square pieces of paper with an area of 4. She cuts them into squares and right triangles in the manner shown in the first diagram. She takes some of the pieces and makes the bird shown in the second diagram. What is the area of the bird?

   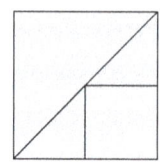 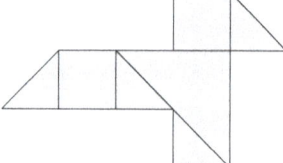

   (A) 3
   (B) 4
   (C) $\frac{9}{2}$
   (D) 5
   (E) 6

7. A bucket was half full. A cleaner added 2 liters to the bucket. The bucket was then three-quarters full. What is the volume of the bucket?

   (A) 10 liters
   (B) 8 liters
   (C) 6 liters
   (D) 4 liters
   (E) 2 liters

8. George built the shape shown using seven unit cubes. How many such cubes does he have to add to make a cube with edges with a length of 3?

   (A) 12
   (B) 14
   (C) 16
   (D) 18
   (E) 20

9. Which of the following calculations gives the largest result?

(A) 44 × 777
(B) 55 × 666
(C) 77 × 444
(D) 88 × 333
(E) 99 × 222

10. The necklace in the picture contains dark gray beads and white beads. Arnold takes one bead after another from the necklace. He always takes a bead from one of the ends. He stops as soon as he has taken the fifth dark gray bead. What is the largest number of white beads that Arnold can take?

(A) 4
(B) 5
(C) 6
(D) 7
(E) 8

4 Points Each

11. Jack has a piano lesson twice a week and Hannah has a piano lesson every other week. In a given quarter, Jack has 15 more lessons than Hannah. How many weeks long is their quarter?

(A) 30
(B) 25
(C) 20
(D) 15
(E) 10

12. In the diagram, the area of each circle is 1 cm². The area common to two overlapping circles is $\frac{1}{8}$ cm². What is the area of the region covered by the five circles shown?

(A) 4 cm²
(B) $\frac{9}{2}$ cm²
(C) $\frac{35}{8}$ cm²
(D) $\frac{39}{8}$ cm²
(E) $\frac{19}{4}$ cm²

13. This year, a grandmother, her daughter, and her granddaughter noticed that the sum of their ages is 100 years. Each of their ages is a power of 2. How old is the granddaughter?

(A) 1
(B) 2
(C) 4
(D) 8
(E) 16

14. Five equal rectangles are placed inside a square with a side length of 24 cm, as shown in the diagram. What is the area of one rectangle?

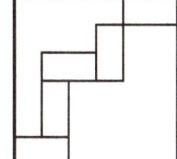

(A) 12 cm²
(B) 16 cm²
(C) 18 cm²
(D) 24 cm²
(E) 32 cm²

15. The heart and the arrow are in the positions shown in the figure. The heart and the arrow start moving at the same time. The arrow moves three places clockwise and the heart moves four places counterclockwise, and then they both stop. They continue the same moves over and over again. After how many such moves will the heart and the arrow land in the same triangular region for the first time?

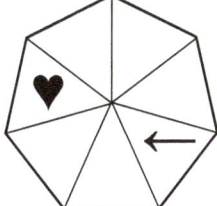

(A) 7
(B) 8
(C) 9
(D) 10
(E) It will never happen.

16 The diagram shows the triangle ABC in which BH is a perpendicular height and AD is the angle bisector at A. The measure of the obtuse angle between BH and AD is four times the measure of angle DAB (see the diagram). What is the measure of angle CAB?

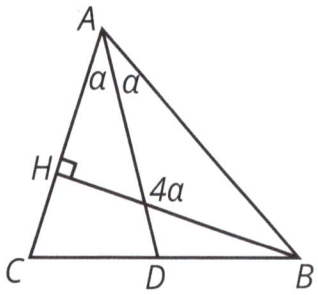

(A) 30°
(B) 45°
(C) 60°
(D) 75°
(E) 90°

17 Six boys share an apartment with two bathrooms which they use every morning beginning at 7:00. There is never more than one person in either bathroom at one time. They spend 8, 10, 12, 17, 21, and 22 minutes at a stretch in the bathroom respectively. What is the earliest time that they can finish using the bathrooms?

(A) 7:45
(B) 7:46
(C) 7:47
(D) 7:48
(E) 7:50

18 A rectangle has sides of length 6 cm and 11 cm. One long side is selected. The bisectors of the angles at either end of that side are drawn. These bisectors divide the other long side into three parts. What are the lengths of these parts?

(A) 1 cm, 9 cm, 1 cm
(B) 2 cm, 7 cm, 2 cm
(C) 3 cm, 5 cm, 3 cm
(D) 4 cm, 3 cm, 4 cm
(E) 5 cm, 1 cm, 5 cm

19 The Pirate Captain Sparrow and his pirate crew dug up some gold coins. They divided the coins among themselves so that each person got the same number of coins. If there had been four pirates less, then each person would have received 10 more coins. However, if there had been 50 fewer coins, then each person would have received 5 coins less. How many coins did they dig up?

(A) 80
(B) 100
(C) 120
(D) 150
(E) 250

20 The average of two positive numbers is 30% less than one of them. By what percentage is the average greater than the other number?

(A) 75%
(B) 70%
(C) 30%
(D) 25%
(E) 20%

5 Points Each

21 Andy enters all the digits from 1 to 9 in the cells of a 3 × 3 table, so that each cell contains one digit. He has already entered 1, 2, 3, and 4, as shown. Two numbers are considered to be "neighbors" if their cells share an edge. After entering all the numbers he notices that the sum of the neighbors of 9 is 15. What is the sum of the neighbors of 8?

(A) 12
(B) 18
(C) 20
(D) 26
(E) 27

22 An antique scale is not working properly. If something is lighter than 1000 g, the scale shows the correct weight. However, if something is heavier than or equal to 1000 g, the scale can show any number above 1000 g. We have 5 weights, each under 1000 g, weighing A g, B g, C g, D g, and E g respectively. When they are weighed in pairs, the scale shows the following: $B + D = 1200$, $C + E = 2100$, $B + E = 800$, $B + C = 900$, $A + E = 700$. Which of the weights is the heaviest?

(A) A
(B) B
(C) C
(D) D
(E) E

23 Quadrilateral $ABCD$ has right angles only at vertices A and D. The numbers show the areas of two of the triangles. What is the area of $ABCD$?

(A) 60
(B) 45
(C) 40
(D) 35
(E) 30

24 Liz and Mary compete in solving problems. Each of them is given the same list of 100 problems. For any problem, the first of them to solve it gets 4 points, and the second to solve it gets 1 point. Liz solved 60 problems, and Mary also solved 60 problems. Together, they got 312 points. How many of the problems were solved by both of them?

(A) 53
(B) 54
(C) 55
(D) 56
(E) 57

25 David rides his bicycle from his home to his friend's house. He was going to arrive at 3:00 p.m., but he spent $\frac{2}{3}$ of the planned time covering $\frac{3}{4}$ of the distance. After that, he rode more slowly and arrived exactly on time. What is the ratio of the speed for the first part of the trip to the speed for the second part?

(A) 5:4
(B) 4:3
(C) 3:2
(D) 2:1
(E) 3:1

26 We have four identical cubes. They are arranged so that a big black circle appears on one face, as shown in the last picture. What can be seen on the opposite face?

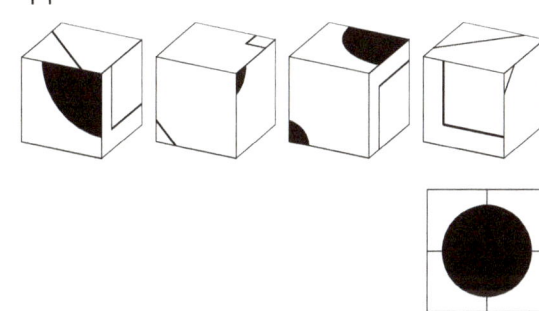

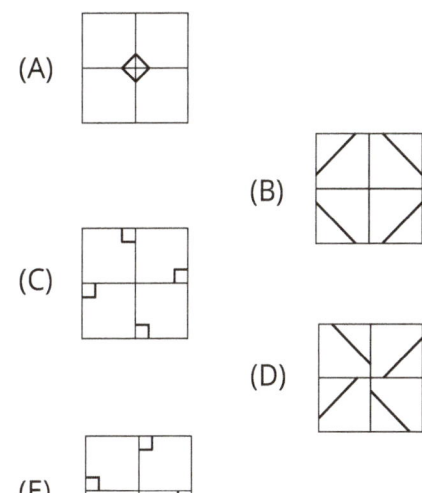

27. A group of 25 people consists of knights, serfs, and damsels. Each knight always tells the truth, each serf always lies, and each damsel alternates between telling the truth and lying. First, 17 of them said "yes" when each was asked, "Are you a knight?" Then, 12 of them said "yes" when each was asked, "Are you a damsel?" Finally, 8 of them said "yes" when each was asked, "Are you a serf?" How many knights are in the group?

(A) 4
(B) 5
(C) 9
(D) 13
(E) 17

28. Several different positive integers are written on the board. Exactly two of them are divisible by 2 and exactly 13 of them are divisible by 13. Let M be the greatest of these numbers. What is the smallest possible value of M?

(A) 169
(B) 260
(C) 273
(D) 299
(E) 325

29. On a pond, there are 16 water lily leaves in a 4 × 4 pattern as shown. A frog sits on a leaf in one of the corners. It then jumps from one leaf to another either horizontally or vertically. The frog always jumps over at least one leaf and never lands on the same leaf twice. What is the greatest number of leaves (including the one it is sitting on) that the frog can reach?

(A) 16
(B) 15
(C) 14
(D) 13
(E) 12

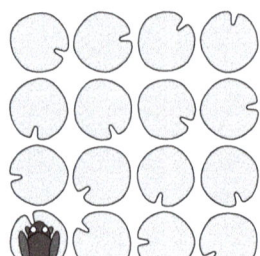

30. A 5 × 5 square is made from 1 × 1 tiles, all with the same pattern, as shown. Any two adjacent tiles have the same color along the shared edge. The perimeter of the large square consists of gray and white segments with a length of 1. What is the smallest possible number of such gray unit segments?

(A) 4
(B) 5
(C) 6
(D) 7
(E) 8

2016

3 Points Each

1. How many whole numbers are there between 20.16 and 3.17?

   (A) 15
   (B) 16
   (C) 17
   (D) 18
   (E) 19

2. Which of the following traffic signs has the largest number of axes of symmetry?

   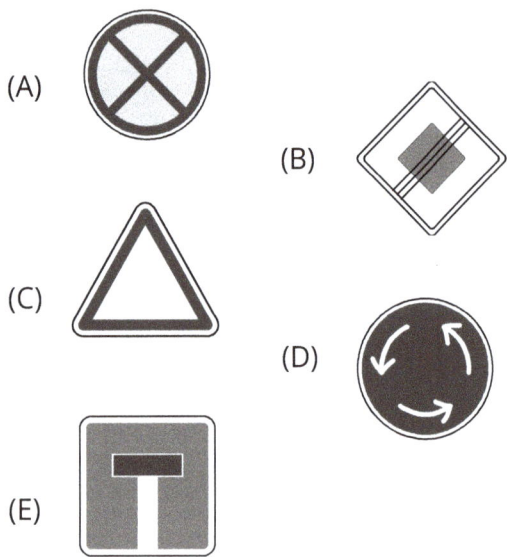

3. What is the sum of the two shaded angles in the figure on the right?

   (A) 150°
   (B) 180°
   (C) 270°
   (D) 320°
   (E) 360°

4. Jenny had to add 26 to a certain number. Instead, she subtracted 26 and obtained −14. What number should she have obtained?

   (A) 28
   (B) 32
   (C) 36
   (D) 38
   (E) 42

5. Joanna turns a card over about its lower edge and then about its right edge, as shown. What does she see?

   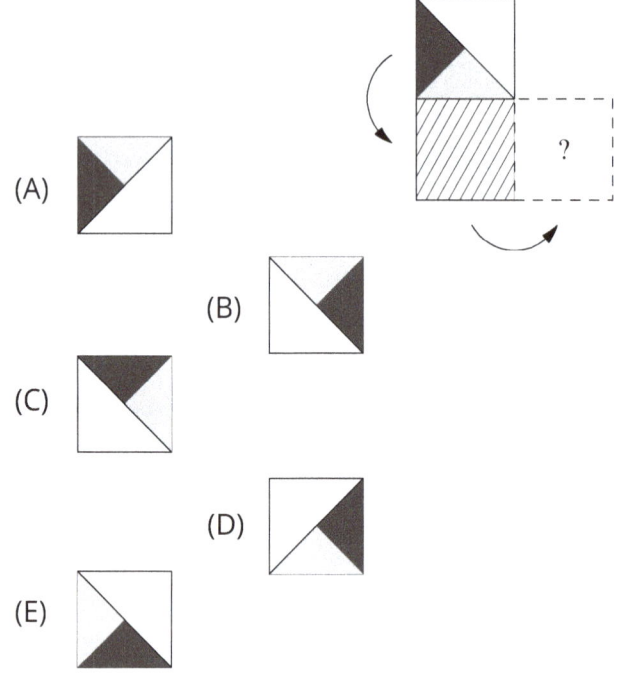

6. Kanga combines 555 groups of 9 stones into a single pile. She then splits the resulting pile into groups of 5 stones. How many groups does she get?

   (A) 999
   (B) 900
   (C) 555
   (D) 111
   (E) 45

7 At my school, 60% of the teachers, which is 45 teachers, bike to school. Only 12% of the teachers use their car to get to school. How many teachers use their car to get to school?

(A) 4
(B) 6
(C) 9
(D) 10
(E) 12

8 What is the size of the shaded area?

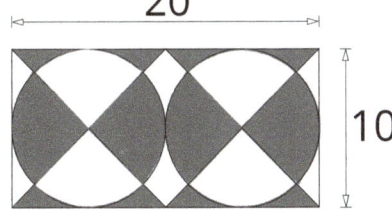

(A) 50
(B) 80
(C) 100
(D) 120
(E) 150

9 One piece of rope has a length of 1 m, and another piece of rope has a length of 2 m. Alex cuts the two pieces of rope into several parts. All the parts have equal lengths. Which of the following cannot be the total number of parts he obtains?

(A) 6
(B) 8
(C) 9
(D) 12
(E) 15

10 Four towns P, Q, R, and S are connected by roads, as shown. A race uses each road exactly once. The race starts at S and finishes at Q. How many possible routes are there for the race?

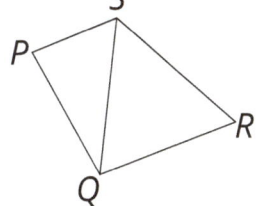

(A) 10
(B) 8
(C) 6
(D) 4
(E) 2

4 Points Each

11 The diagram shows four identical rectangles placed inside a square. The perimeter of each rectangle is 16 cm. What is the perimeter of the big square?

(A) 16 cm
(B) 20 cm
(C) 24 cm
(D) 28 cm
(E) 32 cm

12 Petra has 49 blue beads and one red bead. How many beads must Petra remove so that 90% of her beads are blue?

(A) 4
(B) 10
(C) 29
(D) 39
(E) 40

13 Which of the following fractions has a value closest to $\frac{1}{2}$?

(A) $\frac{25}{79}$
(B) $\frac{27}{59}$
(C) $\frac{29}{57}$
(D) $\frac{52}{79}$
(E) $\frac{57}{92}$

14 Ivor writes down the results of the quarter-finals, the semi-finals, and the final of a knockout tournament. The results are (not necessarily in this order): Bart beat Antony, Carl beat Damien, Glen beat Henry, Glen beat Carl, Carl beat Bart, Ed beat Fred, and Glen beat Ed. Which pair played in the final?

(A) Glen and Henry
(B) Glen and Carl
(C) Carl and Bart
(D) Glen and Ed
(E) Carl and Damien

15. Anne has glued some cubes together, as shown. She rotates the solid to look at it from different angles. Which of the following can she not see?

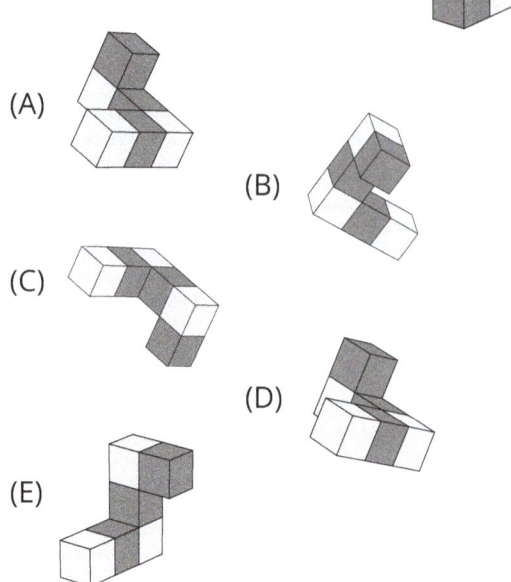

16. Tim, Tom, and Jim are triplets (three brothers born on the same day). Their twin brothers John and James are 3 years younger. Which of the following numbers could be the sum of the ages of the five brothers?

    (A) 36
    (B) 53
    (C) 76
    (D) 89
    (E) 92

17. A 3 cm wide rectangular strip of paper is gray on one side and white on the other. Maria folds the strip, as shown. The gray trapezoids are identical. What is the length of the original strip?

    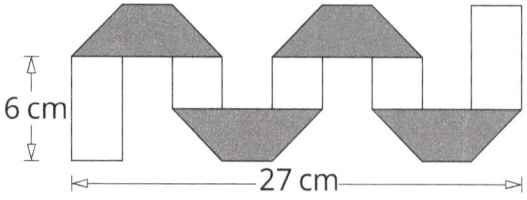

    (A) 36 cm
    (B) 48 cm
    (C) 54 cm
    (D) 57 cm
    (E) 81 cm

18. Two kangaroos, Jum and Per, start to jump at the same time, from the same point, in the same direction. They make one jump per second. Each of Jum's jumps is 6 m in length. Per's first jump is 1 m in length, the second is 2 m, the third is 3 m, and so on. After how many jumps does Per catch up to Jum?

    (A) 10
    (B) 11
    (C) 12
    (D) 13
    (E) 14

19. Seven standard dice are glued together to make the solid shown. The faces of the dice that are glued together have the same number of dots on them. How many dots are on the surface of the solid?

    (A) 24
    (B) 90
    (C) 95
    (D) 105
    (E) 126

20. There are 20 students in a class. They sit in pairs so that exactly one third of the boys sit with a girl, and exactly one half of the girls sit with a boy. How many boys are there in the class?

    (A) 9
    (B) 12
    (C) 15
    (D) 16
    (E) 18

5 Points Each

21 Inside a square with an area of 36, there are shaded regions as shown. The total shaded area is 27. What is $p + q + r + s$?

(A) 4
(B) 6
(C) 8
(D) 9
(E) 10

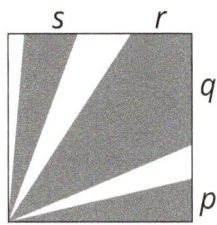

22 Theo's watch is 10 minutes slow, but he believes that it is 5 minutes fast. Leo's watch is 5 minutes fast, but he believes that it is 10 minutes slow. At the same moment, each of them looks at his own watch. Theo thinks it is 12:00. What time does Leo think it is?

(A) 11:30
(B) 11:45
(C) 12:00
(D) 12:30
(E) 12:45

23 Twelve girls met in a café. On average, they ate 1.5 cupcakes each. None of them ate more than two cupcakes, and two of them had only mineral water. How many girls ate two cupcakes?

(A) 2
(B) 5
(C) 6
(D) 7
(E) 8

24 Little Red Riding Hood is delivering waffles to three grannies. She starts with a basket full of waffles. Just before she enters each of the grannies' houses, the Big Bad Wolf eats half of the waffles in her basket. When she leaves the third granny's house, she has no waffles left. She delivers the same number of waffles to each granny. Which of the following numbers definitely divides the number of waffles she started with?

(A) 4
(B) 5
(C) 6
(D) 7
(E) 9

25 The cube below is divided into 64 small cubes. Exactly one of the small cubes is gray. On the first day, the gray cube changes all its neighboring small cubes to gray (two cubes are neighbors if they have a common face). On the second day, all the gray cubes do the same thing. How many gray cubes are there at the end of the second day?

(A) 11
(B) 13
(C) 15
(D) 16
(E) 17

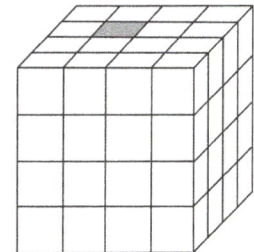

26 Several different positive integers are written on a blackboard. The product of the smallest two of them is 16. The product of the largest two is 225. What is the sum of all the integers?

(A) 38
(B) 42
(C) 44
(D) 58
(E) 243

27 The diagram shows a pentagon. Sepideh draws five circles with centers A, B, C, D, E such that the two circles on each side of the pentagon touch. The lengths of the sides of the pentagon are given. Which point is the center of the largest circle that she draws?

(A) A
(B) B
(C) C
(D) D
(E) E

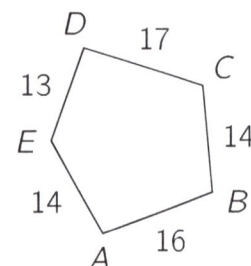

28 Katie writes a different positive integer on each of the fourteen cubes in the pyramid. The sum of the nine integers written on the bottom cubes is equal to 50. The integer written on each of the other cubes is equal to the sum of the integers written on the four cubes underneath it. What is the greatest possible integer that can be written on the top cube?

(A) 80
(B) 98
(C) 104
(D) 110
(E) 118

29 A certain train has five passenger cars, each containing at least one passenger. Two passengers are said to be "neighbors" if either they are in the same car or they are in one of the adjacent cars. Each passenger has either exactly five or exactly ten "neighbors." How many passangers are there on the train?

(A) 13
(B) 15
(C) 17
(D) 20
(E) There is more than one possibility.

30 A 3 × 3 × 3 cube is built from 15 black cubes and 12 white cubes. Five faces of the larger cube are shown.

    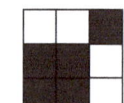

Which of the following is the sixth face of the large cube?

(A)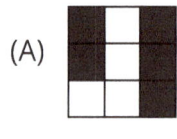

(B)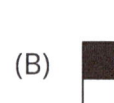

(C)

(D)

(E)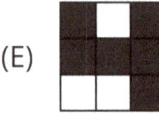

2018

2018

3 Points Each

1 What is the value of (20 + 18) ÷ (20 − 18)?

(A) 18
(B) 19
(C) 20
(D) 34
(E) 36

2 When the letters of the word MAMA are written vertically above one another, the word has a vertical line of symmetry. Which of these words also has a vertical line of symmetry when written in the same way?

(A) ROOT
(B) BOOM
(C) BOOT
(D) LOOT
(E) TOOT

3 A triangle has sides of length 6, 10, and 11. An equilateral triangle has the same perimeter. What is the length of each side of the equilateral triangle?

(A) 6
(B) 9
(C) 10
(D) 11
(E) 27

4 Which number should replace ★ in the equation 2 × 18 × 14 = 6 × ★ × 7 to make it correct?

(A) 8
(B) 9
(C) 10
(D) 12
(E) 15

5 The panels of Fergus' fence are full of holes. One morning, one of the panels fell flat on the ground. Which of the following could Fergus see as he approaches his fence?

(A)

(B)

(C)

(D)

(E)

6 Bertie the Builder is assembling stairs. Each step is 15 cm tall and 15 cm deep, as shown in the diagram. How many steps does he need to assemble in order to reach the second floor which is 3 meters above the first floor in the building?

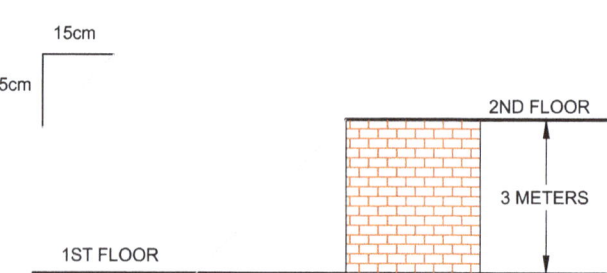

(A) 8
(B) 10
(C) 15
(D) 20
(E) 25

7. A game consists of dropping a ball from the top of the board with rows of pins as shown in the picture. The ball bounces either to the right or to the left each time it hits a pin. One possible route for the ball to take is shown. How many different routes can a ball take to reach bin B?

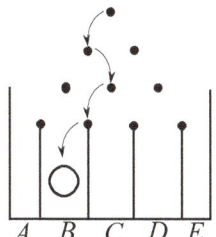

(A) 2
(B) 3
(C) 4
(D) 5
(E) 6

8. A large rectangle is made up of nine identical rectangles whose longer sides are each 10 cm long. What is the perimeter of the large rectangle?

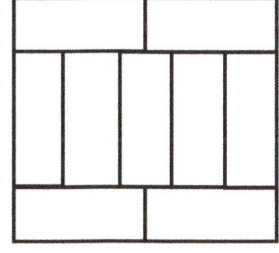

(A) 40 cm
(B) 48 cm
(C) 76 cm
(D) 81 cm
(E) 90 cm

9. The diagram shows a rectangle with dimensions 11 × 7 containing two circles. Each circle touches three of the sides of the rectangle. What is the distance between the centers of the two circles?

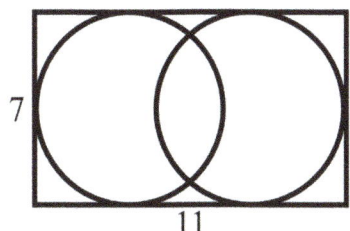

(A) 1
(B) 2
(C) 3
(D) 4
(E) 5

10. Square ABCD has sides with lengths equal to 3 cm. The points M and N lie on AD and AB so that CM and CN split the square into three pieces of the same area. What is the length of DM?

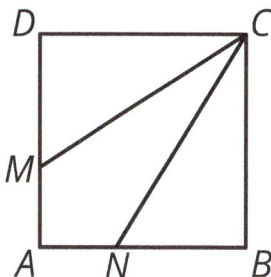

(A) 0.5 cm
(B) 1 cm
(C) 1.5 cm
(D) 2 cm
(E) 2.5 cm

4 Points Each

11. Martha multiplied two 2-digit numbers correctly on a piece of paper. Then she scribbled out three digits as shown. What is the sum of the three digits she scribbled out?

(A) 5
(B) 6
(C) 9
(D) 12
(E) 14

12. A rectangle is divided into 40 identical squares. The rectangle contains more than one row of squares. Andrew found the middle row of squares and colored it in. How many squares did he not color?

(A) 20
(B) 30
(C) 32
(D) 35
(E) 39

13. Philip wants to know the weight of a book with 0.5 gram precision. His weighing scales only weigh in 10 gram increments. What is the smallest number of identical copies of this book that Philip should weigh together to be able to do this?

(A) 5
(B) 10
(C) 15
(D) 20
(E) 50

14. A lion is hidden in one of three rooms. A note on the door of room 1 reads, "The lion is here." A note on the door of room 2 reads, "The lion is not here." A note on the door of room 3 reads, "2 + 3 = 2 × 3." Only one of these sentences is true. In which room is the lion hidden?

(A) room 1
(B) room 2
(C) room 3
(D) It may be in any room.
(E) It may be in either room 1 or room 2.

15. Valeriu draws a zig-zag line inside a rectangle, creating angles of 10°, 14°, 33°, and 26° as shown. What is the measure of angle θ?

(A) 11°
(B) 12°
(C) 16°
(D) 17°
(E) 33°

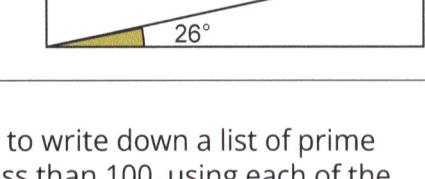

16. Alice wants to write down a list of prime numbers less than 100, using each of the digits 1, 2, 3, 4, and 5 exactly once, and using no other digits. Which prime number must be in her list?

(A) 2
(B) 5
(C) 31
(D) 41
(E) 53

17. A hotel on an island in the Caribbean advertises by using the slogan, "350 days of sun every year!" According to this ad, what is the smallest number of days Willi Burn has to stay at the hotel in 2018 to be certain of having two consecutive days of sun?

(A) 17
(B) 21
(C) 31
(D) 32
(E) 35

18. The diagram shows a rectangle and a line X parallel to its base. Two points A and B lie on X inside the rectangle. The sum of the areas of the two shaded triangles is 10 cm². What is the area of the rectangle?

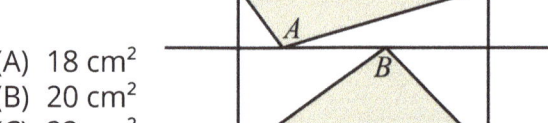

(A) 18 cm²
(B) 20 cm²
(C) 22 cm²
(D) 24 cm²
(E) It depends on the positions of A and B.

19. James wrote a different integer from 1 to 9 in each cell of a 3 × 3 table. He calculated the sum of the integers in each of the rows and in each of the columns of the table. Five of his answers are 12, 13, 15, 16, and 17, in any order. What is his sixth answer?

(A) 17
(B) 16
(C) 15
(D) 14
(E) 13

20. Eleven points are marked from left to right on a straight line. The sum of all the distances between the first point and the other points is 2018. The sum of all the distances between the second point and the other points, including the first one, is 2000. What is the distance between the first and second points?

(A) 1
(B) 2
(C) 3
(D) 4
(E) 5

5 Points Each

21. There are three candidates for one position as class president and 130 students are voting. Suhaimi has 24 votes so far, while Khairul has 29 and Akmal has 37. How many more votes does Akmal need in order to be elected?

(A) 13
(B) 14
(C) 15
(D) 16
(E) 17

22. The diagram shows the net of an unfolded rectangular box. What is the volume of the box?

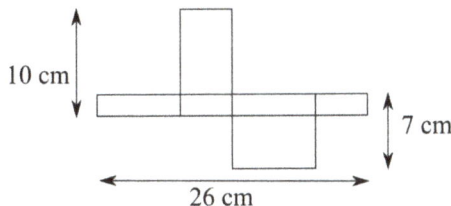

(A) 43 cm³
(B) 70 cm³
(C) 80 cm³
(D) 100 cm³
(E) 1820 cm³

23. Ria wants to write a number in every cell on the border of a 6 × 5 table. In each cell, the number she writes is equal to the sum of the two numbers in the cells with which this cell shares an edge. Two of the numbers are given in the diagram. What number will she write in the cell marked x?

(A) 10
(B) 7
(C) 13
(D) −13
(E) −3

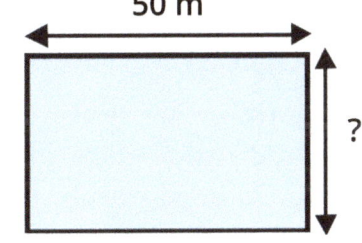

24. Simon and Ian decide to have a race. Simon runs around the perimeter of the pool shown in the diagram while Ian swims lengths of the pool. Simon runs three times faster than Ian swims. Ian swam six lengths of the pool in the same time Simon ran around the pool five times. How wide is the pool?

(A) 25 m
(B) 40 m
(C) 50 m
(D) 80 m
(E) 180 m

25. Freda's flying club designed a flag of a flying dove on graph paper as shown. The area of the dove is 192 cm². All parts of the perimeter of the dove are either parts of a circle or straight lines. What are the dimensions of the flag?

(A) 6 cm × 4 cm
(B) 12 cm × 8 cm
(C) 20 cm × 12 cm
(D) 24 cm × 16 cm
(E) 30 cm × 20 cm

26 Domino tiles are said to be arranged correctly if the number of dots at the ends that touch are the same. Paulius laid six dominoes in a line as shown in the diagram. He can make a move by either swapping the position of any two dominoes or by rotating one domino. What is the smallest number of moves he needs to make to arrange all the tiles correctly?

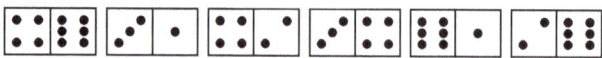

(A) 1
(B) 2
(C) 3
(D) 4
(E) It is impossible to do.

27 Points N, M, and L lie on the sides of the equilateral triangle ABC, such that NM ⊥ BC, ML ⊥ AB, and LN ⊥ AC as shown in the diagram. The area of triangle ABC is 36. What is the area of triangle LMN?

(A) 9
(B) 12
(C) 15
(D) 16
(E) 18

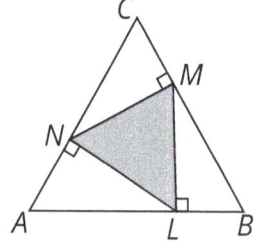

28 Azmi, Burhan, and Choo went shopping. Burhan spent only 15% of what Choo spent. However, Azmi spent 60% more than Choo. Together they spent $55. How many dollars did Azmi spend?

(A) 3
(B) 20
(C) 25
(D) 26
(E) 32

29 Viola is practicing the long jump. The average distance she has jumped so far today is 3.80 m. On her next jump, she jumped 3.99 m and her average increased to 3.81 m. What distance must she jump with her next jump to increase her average to 3.82 m?

(A) 3.97 m
(B) 4.00 m
(C) 4.01 m
(D) 4.03 m
(E) 4.04 m

30 In isosceles triangle ABC, points K and L are marked on sides AB and BC respectively so that AK = KL = LB and KB = AC. What is the measure of angle ABC?

(A) 30°
(B) 35°
(C) 36°
(D) 40°
(E) 44°

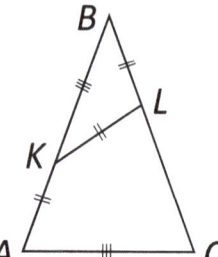

2020

3 Points Each

1 How many of the following four numbers, 2, 20, 202, and 2020, are prime?

(A) 0
(B) 1
(C) 2
(D) 3
(E) 4

2 In which of the regular polygons is the marked angle the largest?

(A)

(B)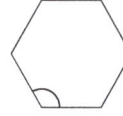

(C)

(D)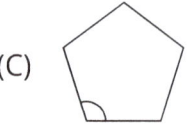

(E)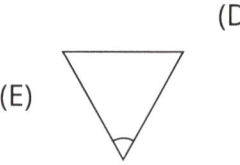

3 Miguel solves six Olympiad problems every day and Lazaro solves four Olympiad problems every day. How many days does it take Lazaro to solve the same number of problems as Miguel solves in four days?

(A) 4
(B) 5
(C) 6
(D) 7
(E) 8

4 Which of these fractions has the largest value?

(A) $\dfrac{8+5}{3}$

(B) $\dfrac{8}{3+5}$

(C) $\dfrac{3+5}{8}$

(D) $\dfrac{8+3}{5}$

(E) $\dfrac{3}{8+5}$

5 A large square is divided into smaller squares. In one of the squares, a diagonal is also drawn. What fraction of the large square is shaded?

(A) $\dfrac{4}{5}$
(B) $\dfrac{3}{8}$
(C) $\dfrac{4}{9}$
(D) $\dfrac{1}{3}$
(E) $\dfrac{1}{2}$

6 There are 4 teams in a soccer tournament. Each team plays each of the other teams exactly once. In each match, the winner receives 3 points and the loser receives 0 points. In the case of a tie, both teams receive 1 point. After all the matches have been played, which of the following total number of points is it impossible for any team to have received?

(A) 4
(B) 5
(C) 6
(D) 7
(E) 8

7 The diagram shows a shape made up of 36 identical small triangles. What is the smallest number of such triangles that can be added to the shape to turn it into a hexagon?

(A) 10
(B) 12
(C) 15
(D) 18
(E) 24

8. Kanga wants to multiply together three different numbers from the following list: −5, −3, −1, 2, 4, and 6. What is the smallest result she can obtain?

(A) −200
(B) −120
(C) −90
(D) −48
(E) −15

9. If John goes to school by bus and walks back, he travels for 3 hours. If he goes by bus both ways, he travels for 1 hour. How long does it take him if he walks both ways?

(A) 3.5 hours
(B) 4 hours
(C) 4.5 hours
(D) 5 hours
(E) 5.5 hours

10. A number is written in each cell of a 3 × 3 square. Unfortunately, the numbers are not visible because they are covered in ink. However, the sum of the numbers in each row and the sum of the numbers in two of the columns are all known, as shown by the arrows in the diagram. What is the sum of the numbers in the third column?

(A) 41
(B) 43
(C) 44
(D) 45
(E) 47

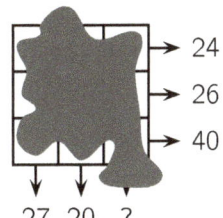

4 Points Each

11. The shortest path from Atown to Cetown runs through Betown. The two signposts shown are set up along this path. What distance was written on the broken sign?

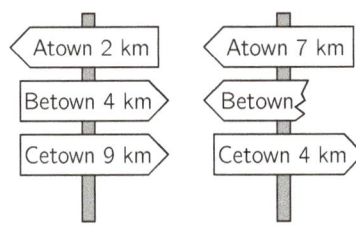

(A) 1 km
(B) 3 km
(C) 4 km
(D) 5 km
(E) 9 km

12. Anna wants to walk 5 km on average each day in March. At bedtime on March 16th, she realized that she had walked 95 km so far. What distance does she need to walk on average for the remaining days of the month to achieve her target?

(A) 5.4 km
(B) 5 km
(C) 4 km
(D) 3.6 km
(E) 3.1 km

13. Which of the following shows what you would see when the object in the diagram is viewed from above?

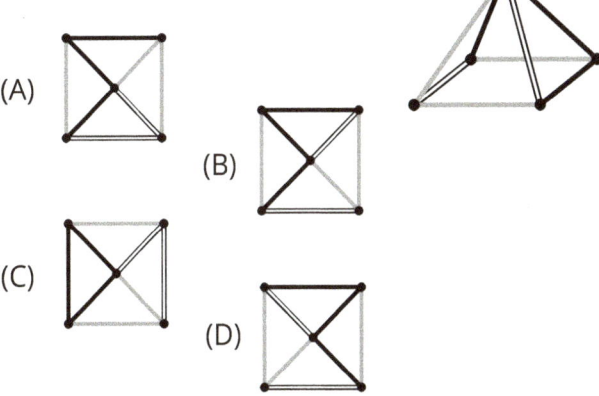

14 Every student in a class swims, dances, or does both. Three-fifths of the class swim and three-fifths dance. Five students both swim and dance. How many students are in the class?

(A) 15
(B) 20
(C) 25
(D) 30
(E) 35

15 Sacha's garden has the shape shown. All the sides are either parallel or perpendicular to each other. Some of the dimensions are shown in the diagram. What is the perimeter of Sacha's garden?

(A) 22
(B) 23
(C) 24
(D) 25
(E) 26

16 Andrew buys 27 identical small cubes, each with two adjacent faces painted red. He then uses all of these cubes to build a large cube. What is the largest number of completely red faces of the large cube that he can make?

(A) 2
(B) 3
(C) 4
(D) 5
(E) 6

17 A large square consists of four identical rectangles and a small square. The area of the large square is 49 cm² and the length of the diagonal AB of one of the rectangles is 5 cm. What is the area of the small square?

(A) 1 cm²
(B) 4 cm²
(C) 9 cm²
(D) 16 cm²
(E) 25 cm²

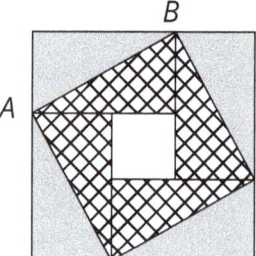

18 Werner's salary is 20% of his boss's salary. By what percentage should Werner's salary increase to become equal to his boss's salary?

(A) 80%
(B) 120%
(C) 180%
(D) 400%
(E) 520%

19 Irene made a "city" using identical wooden cubes. One of the diagrams shows the view from above the "city" and the other the view from one of the sides. However, it is not known from which side the side view was taken. What is the largest number of cubes that Irene could have used?

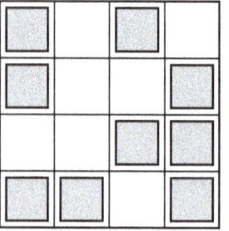

(A) 25
(B) 24
(C) 23
(D) 22
(E) 21

20 Aisha has a strip of paper with the numbers 1, 2, 3, 4, and 5 written in five cells as shown. She folds the strip so that the cells overlap, forming 5 layers.

Which of the following configurations, from the top layer to the bottom layer, is it not possible to obtain?

(A) 3, 5, 4, 2, 1
(B) 3, 4, 5, 1, 2
(C) 3, 2, 1, 4, 5
(D) 3, 1, 2, 4, 5
(E) 3, 4, 2, 1, 5

5 Points Each

21 Twelve colored cubes are arranged in a row. There are 3 blue cubes, 2 yellow cubes, 3 red cubes, and 4 green cubes, but not in that order. There is a yellow cube at one end and a red cube at the other end. The red cubes are all touching. The green cubes are also all touching. The tenth cube from the left is blue. What color is the sixth cube from the left?

(A) green
(B) yellow
(C) blue
(D) red
(E) red or blue

22 Zaida took a square piece of paper and folded two of its sides to the diagonal, as shown, to obtain a quadrilateral. What is the size of the largest angle of the quadrilateral?

(A) 112.5°
(B) 120°
(C) 125°
(D) 135°
(E) 150°

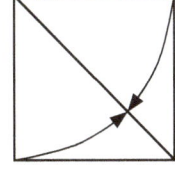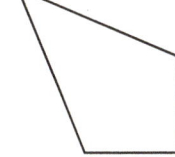

23 How many four-digit numbers A are there such that half of the number A is divisible by 2, a third of A is divisible by 3, and a fifth of A is divisible by 5?

(A) 1
(B) 7
(C) 9
(D) 10
(E) 11

24 In the finals of the dancing competition, each of the three members of the jury gives the five competitors 0 points, 1 point, 2 points, 3 points, or 4 points. No two competitors get the same mark from any individual judge. Adam knows all the sums of the marks and a few single marks, as shown. How many points did Adam get from judge III?

	Adam	Berta	Clara	David	Emil
I	2	0			
II		2	0		
III					
Sum	7	5	3	4	11

(A) 0
(B) 1
(C) 2
(D) 3
(E) 4

25 Saniya writes a positive integer on each edge of a square. She also writes at each vertex the product of the numbers on the two edges that meet at that vertex. The sum of the numbers at the vertices is 15. What is the sum of the numbers on the edges of the square?

(A) 6
(B) 7
(C) 8
(D) 10
(E) 15

26 Sophia has 52 identical isosceles right triangles. She wants to make a square using some of them. How many different sized squares can she make?

(A) 6
(B) 7
(C) 8
(D) 9
(E) 10

---

27 Cleo builds a pyramid with metal spheres. The square base consists of 4 × 4 spheres as shown in the figure. The floors consist of 3 × 3 spheres, 2 × 2 spheres, and a final sphere at the top. At each point of contact between two spheres, a blob of glue is placed. How many blobs of glue will Cleo place?

(A) 72
(B) 85
(C) 88
(D) 92
(E) 96

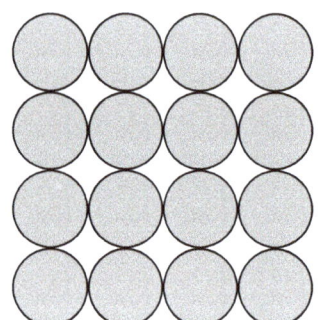

---

28 Four children are in the four corners of a 10 m × 25 m pool. Their coach is standing somewhere on one side of the pool. When he calls them, three children get out and walk as short a distance as possible along the edges of the pool to meet him. They walk 50 m in total. What is the shortest distance the trainer needs to walk to get to the fourth child?

(A) 10 m
(B) 12 m
(C) 15 m
(D) 20 m
(E) 25 m

---

29 Anne, Boris, and Carl ran a race. They started at the same time, and their speeds were constant. When Anne finished, Boris had 15 m to run, and Carl had 35 m to run. When Boris finished, Carl had 22 m to run. What is the distance they ran?

(A) 135 m
(B) 140 m
(C) 150 m
(D) 165 m
(E) 175 m

---

30 The statements give clues to the identity of a four-digit number.

Two digits are correct but in the wrong places:
| 4 | 1 | 3 | 2 |

One digit is correct and in the right place:
| 9 | 8 | 2 | 6 |

Two digits are correct, with one of them being in the right place and the other one in the wrong place:
| 5 | 0 | 7 | 9 |

One digit is correct but in the wrong place:
| 2 | 7 | 4 | 1 |

None of the digits are correct:
| 7 | 6 | 4 | 2 |

What is the last digit of the four-digit number?

(A) 0
(B) 1
(C) 3
(D) 5
(E) 9

2022

2022

3 Points Each

1 Sanath paddled around five buoys, as shown. Around which of the buoys did Sanath paddle in a clockwise direction?

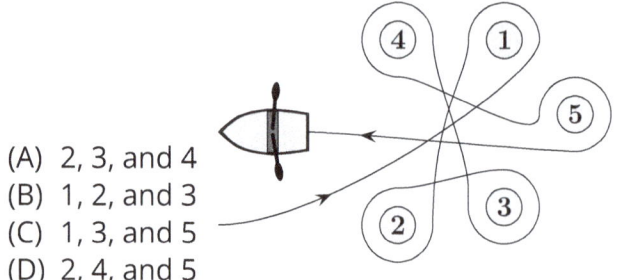

(A) 2, 3, and 4
(B) 1, 2, and 3
(C) 1, 3, and 5
(D) 2, 4, and 5
(E) 2, 3, and 5

2 Beate rearranges the five numbered pieces shown to display the smallest possible nine-digit number. Which piece does she place at the right-hand end?

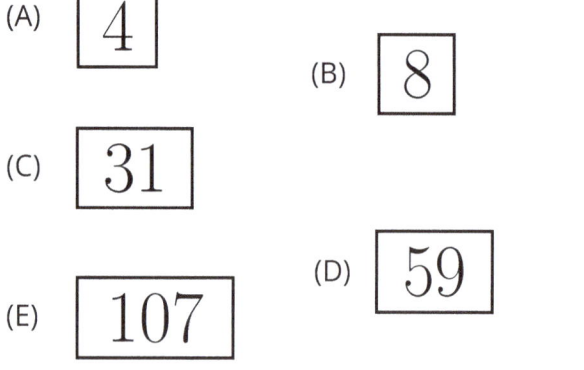

(A) 4
(B) 8
(C) 31
(D) 59
(E) 107

3 Kengu enjoys jumping on the number line. He always makes two large jumps followed by three small jumps, as shown, and then repeats this process over and over again. Kengu starts his jumping routine on 0. On which of these numbers will Kengu land during his routine?

(A) 82
(B) 83
(C) 84
(D) 85
(E) 86

4 The license plate of Kangy's car fell off. He put it back upside down, but luckily this didn't make any difference. Which of the following could be Kangy's license plate?

(A) 04 NSN 40
(B) 60 HOH 09
(C) 80 BNB 08
(D) 03 HNH 30
(E) 08 XBX 80

5 Rob the Builder has a brick whose shortest side is 4 cm. He uses several such bricks to build the cube shown. What are the dimensions, in cm, of his brick?

(A) 4 × 6 × 12
(B) 4 × 6 × 16
(C) 4 × 8 × 12
(D) 4 × 8 × 16
(E) 4 × 12 × 16

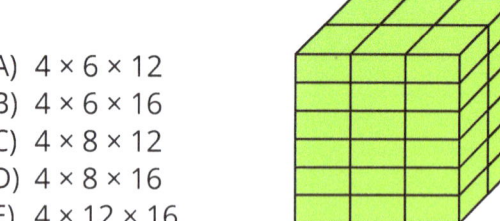

6 The black and white caterpillar shown in the picture curls up to sleep. Which of the following could be seen?

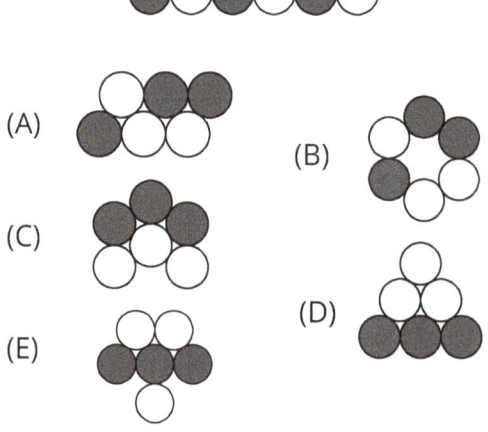

7. In the statement, there are five empty spaces. Sanja wants to fill four of them with plus signs and one with a minus sign so that the statement is correct. Where should she place the minus sign?

6 ☐ 9 ☐ 12 ☐ 15 ☐ 18 ☐ 21 = 45

(A) between 6 and 9
(B) between 9 and 12
(C) between 12 and 15
(D) between 15 and 18
(E) between 18 and 21

8. There are five big trees and three paths in a park. In which region of the park should a new tree be planted so that for each path there is the same number of trees on both sides?

(A) A
(B) B
(C) C
(D) D
(E) E

9. How many positive integers between 100 and 300 have only odd digits?

(A) 25
(B) 50
(C) 75
(D) 100
(E) 150

10. Gerard wrote down the sum of squares of two numbers, as shown. Unfortunately, some of the digits cannot be seen because they are covered in ink. What is the last digit of the first number?

$(2?)^2 + (12)^2 = 7133029$

(A) 3
(B) 4
(C) 5
(D) 6
(E) 7

4 Points Each

11. The distance between two shelves in the cupboard in Monica's kitchen is 36 cm. She knows that a stack of 8 of her favorite glasses is 42 cm tall and that a stack of 2 glasses is 18 cm tall. What is the largest number of glasses that can be stacked and still fit between two shelves?

(A) 3
(B) 4
(C) 5
(D) 6
(E) 7

12. On a standard die, the sum of the numbers of dots on opposite faces is always 7. Four standard dice are glued together, as shown. What is the minimum number of dots that could lie on the whole surface?

(A) 52
(B) 54
(C) 56
(D) 58
(E) 60

13 Three sisters, whose average age is 10, all have different ages. When they get together in pairs, the average ages of two such pairs are 11 and 12. What is the age of the eldest sister?

(A) 10
(B) 11
(C) 12
(D) 14
(E) 16

14 Tony the Gardener planted tulips and daisies in a square flowerbed with a side length of 12 m, arranged as shown. What is the total area of the regions in which he planted daisies?

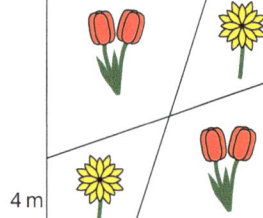

(A) 48 m²
(B) 46 m²
(C) 44 m²
(D) 40 m²
(E) 36 m²

15 In my office, there are two clocks. One clock gains one minute every hour and the other loses two minutes every hour. Yesterday I set them both to the correct time, but when I looked at them today, I saw that the time shown on one was 11:00 a.m. and shown on the other was 12:00 noon. What time was it when I set the two clocks?

(A) 11:00 p.m.
(B) 7:40 p.m.
(C) 3:40 p.m.
(D) 2:00 p.m.
(E) 11:20 a.m.

16 Werner wrote several positive numbers smaller than 7 on a piece of paper. Ria then crossed out all of Werner's numbers and replaced each of them with their difference from 7. The sum of Werner's numbers was 22. The sum of Ria's numbers is 34. How many numbers did Werner write down?

(A) 7
(B) 8
(C) 9
(D) 10
(E) 11

17 The numbers 1 to 8 are placed, once each, in the circles shown. The numbers by the arrows show the products of the three numbers in the circles on that straight line. What is the sum of the numbers in the three circles at the bottom of the figure?

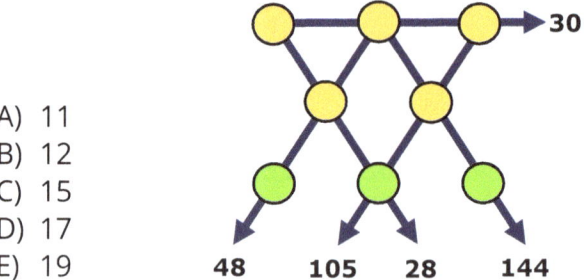

(A) 11
(B) 12
(C) 15
(D) 17
(E) 19

18 The area of the intersection of a circle and a triangle is 45% of the area of their union. The area of the triangle outside the circle is 40% of the area of their union. What percentage of the circle lies outside the triangle?

(A) 20%
(B) 25%
(C) 30%
(D) 35%
(E) 50%

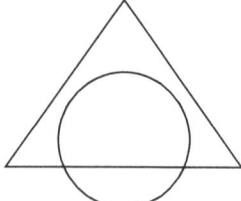

19 In how many ways can the following shape be completely covered with nine tiles, such as those on its right side?

(A) 1
(B) 6
(C) 8
(D) 9
(E) 12

20 Marc always bikes at the same speed and he always walks at the same speed. He can cover the round trip from his home to school and back again in 20 minutes when he bikes and in 60 minutes when he walks. Yesterday Marc started biking to school but stopped and left his bike at Eva's house on the way before finishing his trip to school on foot. On the way back, he walked to Eva's house, picked up his bike, and then biked the rest of the way home. His total travel time was 52 minutes. What fraction of his trip did Marc make by bike?

(A) $\frac{1}{6}$
(B) $\frac{1}{5}$
(C) $\frac{1}{4}$
(D) $\frac{1}{3}$
(E) $\frac{1}{2}$

5 Points Each

21 Jenny decided to enter numbers into the cells of a 3 × 3 table so that the sum of the numbers in all four possible 2 × 2 squares will be the same. The numbers in three of the corner cells have already been written, as shown. Which number should she write in the fourth corner cell?

(A) 0
(B) 1
(C) 4
(D) 5
(E) 6

22 The villages A, B, C, and D are situated, not necessarily in that order, on a long straight road. The distance from A to C is 75 km, the distance from B to D is 45 km, and the distance from B to C is 20 km. Which of the following cannot be the distance from A to D?

(A) 10 km
(B) 50 km
(C) 80 km
(D) 100 km
(E) 140 km

23 The large rectangle ABCD is divided into seven identical rectangles. What is the ratio AB/BC?

(A) $\frac{1}{2}$
(B) $\frac{4}{3}$
(C) $\frac{8}{5}$
(D) $\frac{12}{7}$
(E) $\frac{7}{3}$

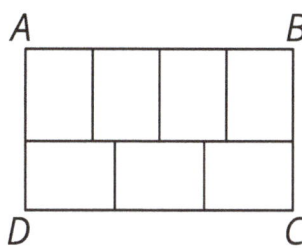

24 A painter wanted to mix 2 liters of blue paint with 3 liters of yellow paint to make 5 liters of green paint. However, by mistake, he used 3 liters of blue and 2 liters of yellow so that he made the wrong shade of green. What is the smallest amount of this green paint that he must throw away so that, using the rest of his green paint and some extra blue and/or yellow paint, he could make 5 liters of paint of the correct shade of green?

(A) $\frac{5}{3}$ liters
(B) $\frac{3}{2}$ liters
(C) $\frac{2}{3}$ liters
(D) $\frac{3}{5}$ liters
(E) $\frac{5}{9}$ liters

25. A builder has two identical bricks. She places them side by side in three different ways, as shown. The surface areas of the three shapes obtained are 72, 96, and 102. What is the surface area of the original brick?

(A) 36
(B) 48
(C) 52
(D) 54
(E) 60

26. What is the smallest number of cells that need to be colored in a 5 × 5 square so that any 1 × 4 or 4 × 1 rectangle lying inside the square has at least one cell colored?

(A) 5
(B) 6
(C) 7
(D) 8
(E) 9

27. Mowgli asks a zebra and a panther what day it is. The zebra always lies on Monday, Tuesday, and Wednesday, and always tells the truth on the other days. The panther always lies on Thursday, Friday, and Saturday, and always tells the truth on the other days. The zebra says, "Yesterday was one of my lying days." The panther says, "Yesterday was also one of my lying days." What day is it?

(A) Thursday
(B) Friday
(C) Saturday
(D) Sunday
(E) Monday

28. Several points were marked on a line. Renard then marked another point between each pair of adjacent points on the line. He repeated this process a further three times. There are now 225 points marked on the line. How many points were marked on the line initially?

(A) 10
(B) 12
(C) 15
(D) 16
(E) 25

29. An isosceles triangle ABC, with AB = AC, is split into three smaller isosceles triangles, as shown, so that AD = DB, CE = CD, and BE = EC. (Note that the diagram is not drawn to scale.) What is the size, in degrees, of angle BAC?

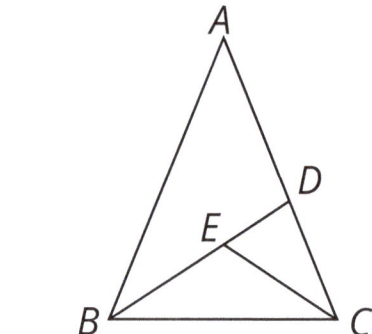

(A) 24
(B) 28
(C) 30
(D) 35
(E) 36

30. There are 2022 kangaroos and some koalas living across seven parks. In each park, the number of kangaroos is equal to the total number of koalas in all the other parks. How many koalas live in the seven parks in total?

(A) 288
(B) 337
(C) 576
(D) 674
(E) 2022

2024

2024

3 Points Each

1. Which of the following strings cannot be transformed into the string on the right without cutting?

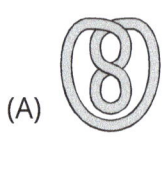
(A)

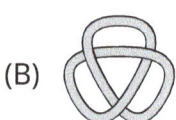
(B)

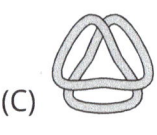
(C)

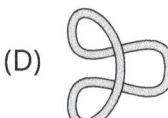
(D)

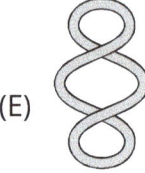
(E)

2. A shape is made of equal-sized pentagonal tiles. Which of the following tiles can be placed in the empty space to produce two closed curves?

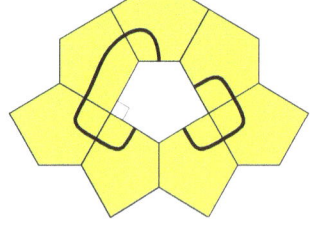

(A)

(B)

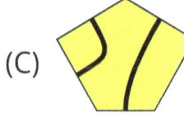
(C)

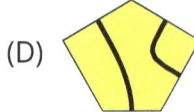
(D)

(E)

3. The first diagram shows a rhombus. The area of the first diagram is increased by adding two right triangles, as shown. By what percentage has the area increased?

(A) 20%
(B) 25%
(C) 30%
(D) 40%
(E) 50%

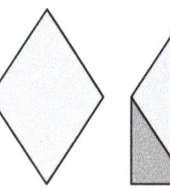

4. What is the value of $\dfrac{20 \times 24}{2 \times 0 + 2 \times 4}$?

(A) 12
(B) 30
(C) 48
(D) 60
(E) 120

5. Julia cuts off the four vertices of a regular tetrahedron, as shown. How many vertices does the shape that remains have?

(A) 8
(B) 9
(C) 11
(D) 12
(E) 15

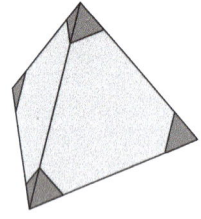

6. Ria has three counters marked 1, 5, and 11, as shown. She wants to place them side by side to make a four-digit number. How many different four-digit numbers can she make?

(A) 3
(B) 4
(C) 6
(D) 8
(E) 9

7 A fruit bowl contains five types of fruit:

The fruit is shared so that everyone gets a different type of fruit and everyone gets a type of fruit that they like.

Who gets  ?

(A) Al
(B) Bok
(C) Cam
(D) Don
(E) Eva

---

8 The weight restriction notice for an elevator says it can carry either 12 adults or 20 children. According to the weight restriction, what is the largest number of children that can ride in the elevator with nine adults?

(A) 3
(B) 4
(C) 5
(D) 6
(E) 8

---

9 Four different positive integers are placed on a grid and then covered up. The products of the integers in each row and in each column are shown in the diagram. What is the sum of the four integers?

(A) 10
(B) 12
(C) 13
(D) 14
(E) 15

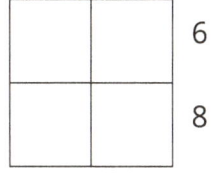

---

10 The length of a set of four well-parked supermarket carts which fit into each other is 108 cm. The length of a set of ten well-parked supermarket carts is 168 cm. What is the length of a single supermarket cart?

(A) 60 cm
(B) 68 cm
(C) 78 cm
(D) 88 cm
(E) 90 cm

4 Points Each

11 Carolina baked a cake and cut it into ten equal pieces. She ate one piece and then arranged the remaining pieces evenly, as shown. What is the measure of the angle between any two adjacent pieces?

(A) 5°
(B) 4°
(C) 3°
(D) 2°
(E) 1°

12. Justin can make a 4 × 4 square, where the sum of the numbers in all four rows and all four columns is the same, from the three pieces shown:

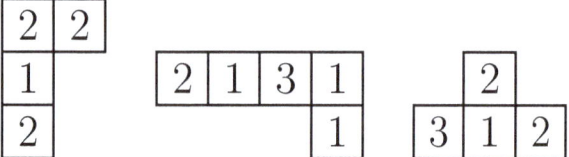

and one additional piece. Which of the following pieces is needed to complete his square?

(A) | 1 | 1 | 3 |

(B) | 2 | 1 | 0 |

(C) | 1 | 2 | 1 |

(D) | 2 | 2 | 2 |

(E) | 2 | 2 | 3 |

13. A square has a side length of 10 m. It is divided into parts by three straight line segments, as shown. The areas of the two shaded triangles are A and B. What is the value of A − B?

(A) 0 m²
(B) 1 m²
(C) 2 m²
(D) 5 m²
(E) 10 m²

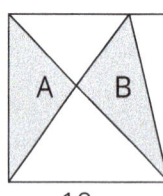

10 m

14. Paula the penguin goes fishing every day and always brings back twelve fish for her two chicks. Each day, she gives the first chick she sees seven fish and gives the second chick five fish, which they eat. In the last few days one chick has eaten 44 fish. How many has the other chick eaten?

(A) 34
(B) 40
(C) 46
(D) 52
(E) 58

15. John had a large number of identical cubes. He made the structure shown below by taking a single cube and then sticking another cube to each face. He wants to make an extended structure in the same way so that each face of his original structure will have a cube stuck to it. How many extra cubes will he need to complete his extended structure?

(A) 18
(B) 16
(C) 14
(D) 12
(E) 10

16. A kangaroo jumps up a mountain and then jumps back down along the same route. It covers three times the distance with each downhill jump as it does with each uphill jump. Going uphill, it covers 1 meter per jump. In total, the kangaroo makes 2024 jumps. What is the total distance, in meters, that the kangaroo jumps?

(A) 506
(B) 1012
(C) 2024
(D) 3036
(E) 4048

17. Gerard cuts a large rectangle into four smaller rectangles. The perimeters of three of these smaller rectangles are 16, 18, and 24, as shown in the diagram. What is the perimeter of the fourth small rectangle?

(A) 8
(B) 10
(C) 12
(D) 14
(E) 16

18. Water makes up 80 percent of the mass of fresh mushrooms. However, water makes up only 20 percent of the mass of dried mushrooms. By what percentage does the mass of a mushroom decrease during drying?

(A) 60
(B) 70
(C) 75
(D) 80
(E) 85

19. Therese the tiler is planning to make a large square mosaic floor with a repeating pattern. She is using hexagonal and triangular tiles arranged as shown in the diagram. She thinks she will use 3000 hexagonal tiles to make the whole floor. Approximately how many triangular tiles will she need?

(A) 1000
(B) 1500
(C) 3000
(D) 6000
(E) 9000

20. Nine cards numbered from 1 to 9 were placed on the table. Alexa, Bart, Clara, and Diana each picked up two of the cards. Alexa said, "My numbers add up to 6." Bart said, "The difference between my numbers is 5." Clara said, "The product of my numbers is 18." Diana said, "One of my numbers is twice the other one." All four made a true statement. Which number was left on the table?

(A) 1
(B) 3
(C) 6
(D) 8
(E) 9

5 Points Each

21. The digits 0 – 9 can be drawn with horizontal and vertical segments, as shown.

0123456789

Greg chooses three different digits. In total, his digits have 5 horizontal segments and 10 vertical segments. What is the sum of his three digits?

(A) 9
(B) 10
(C) 14
(D) 18
(E) 19

22. Thomas wants to shade two additional squares on the diagram shown so that the resulting pattern has a single axis of symmetry. In how many different ways can he do this?

(A) 2
(B) 3
(C) 4
(D) 5
(E) 6

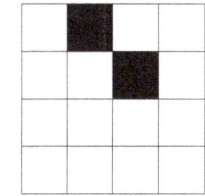

23. The diagram shows three semicircles inside a rectangle. The middle semicircle touches the other two semicircles which, in turn, each touch a shorter side of the rectangle. The largest semicircle also touches one of the longer sides of the rectangle. The shortest distances from that side of the rectangle to the other two semicircles are 5 cm and 7 cm respectively, as shown. What is the perimeter, in cm, of the rectangle?

(A) 82
(B) 92
(C) 96
(D) 108
(E) 120

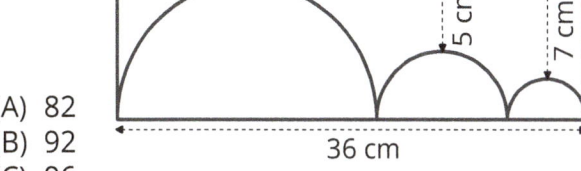

24. A group of 50 students are sitting in a circle and throwing a ball. Each student who gets the ball throws it to the 6th student sitting counterclockwise from where they are sitting, who catches it. Freda catches the ball 100 times. In that time, how many students never get to catch the ball?

(A) 0
(B) 8
(C) 10
(D) 25
(E) 40

25. Daniel wants to complete the diagram so that each box in the middle and top rows will contain the product of the values in the two boxes below it and each box contains a positive integer. He wants the value in the top box to be 720. How many different values can the integer n take?

(A) 1
(B) 4
(C) 5
(D) 6
(E) 8

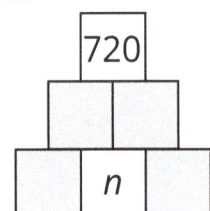

26. Farmer Frank is selling chicken and duck eggs. He has baskets holding 4, 6, 12, 13, 22, and 29 eggs. His first customer buys all the eggs in one basket. Frank notices that the number of chicken eggs he has left is twice the number of duck eggs. How many eggs did the customer buy?

(A) 4
(B) 12
(C) 13
(D) 22
(E) 29

27. Three angles α, β, and γ are marked on graph paper, as shown. What is the value of $\alpha + \beta + \gamma$?

(A) 60°
(B) 70°
(C) 75°
(D) 90°
(E) 120°

28. Captain Flint asked four of his pirates to write on a piece of paper how many gold, silver, and bronze coins were in the treasure chest. Their responses are shown in the diagram but unfortunately part of the paper was damaged. Only one of the four pirates told the truth. The other three lied in all their answers. The total number of coins is 30. Who told the truth?

(A) Tom
(B) Al
(C) Pit
(D) Jim
(E) We cannot be sure.

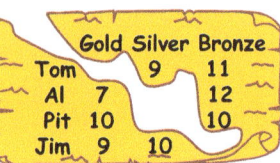

29. Alex drives from point A to point B, then immediately returns to A. Bob drives from point B to point A, then immediately returns to B. They travel on the same road, start at the same time, and each travels at a constant speed. Alex's speed is three times Bob's speed. They pass each other for the first time 15 minutes after the start. How long after the start will they pass each other for the second time?

(A) 20 min
(B) 25 min
(C) 30 min
(D) 35 min
(E) 45 min

30. In the pentagon $ABCDE$, $\angle A = \angle B = 90°$, $AE = BC$, and $ED = DC$. Four points are marked on AB dividing it into five equal parts. Then perpendiculars are drawn through these points, as shown in the diagram. The dark shaded region has an area of 13 cm² and the light shaded region has an area of 10 cm². What is the area, in cm², of the entire pentagon?

(A) 45
(B) 47
(C) 49
(D) 58
(E) 60

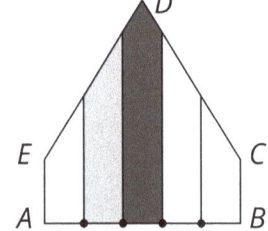

Part II
Solutions

2006

3 Point Solutions

1 **(B) 16th**
The competition took place in the following years: 1991, 1992, 1993, 1994, 1995, 1996, 1997, 1998, 1999, 2000, 2001, 2002, 2003, 2004, and 2005, a total of 15 times. In the year 2006 the competition will take place for the 16th time.

2 **(D) 126**
20 × (0 + 6) − (20 × 0) + 6 = 20 × 6 − 0 + 6 = = 120 + 6 = 126

3 **(D) 30%**
Draw a line from O to the bottom right vertex. The triangle on the left is 10% of the pentagon, and the triangle on the right is 20% of the pentagon. Thus it is a total of 30%.

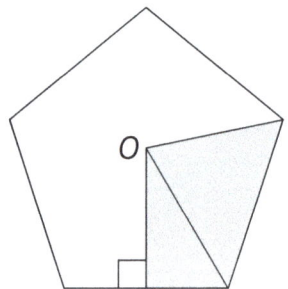

4 **(E) 260**
The sum of the lengths of any two sides of a triangle must be greater than the third side (Triangle Inequality Theorem). 120 + 130 < 260. Therefore, of the options listed, 260 cannot be the third side of the triangle.

5 **(D) 700**
1200 + 1500 + 6 − 2006 = 700

6 **(B) 58**
The base of the small cube and the area on the larger cube which it covers are not part of the surface of the solid.
6 × 9 + 6 × 1 − 2 × 1 = 58.

7 **(A)** $\frac{1}{20}$ **liter**
$\frac{3}{4}$ of $\frac{1}{3}$ liter − $\frac{1}{5}$ liter = $\frac{1}{4}$ liter − $\frac{1}{5}$ liter = $\frac{1}{20}$ liter

8 **(D) 27**
For any isosceles triangle with legs equal to 7, the longest length of the base expressed with whole number is 13, as it must be smaller than the sum of other sides. Thus sum of all sides is 7 + 7 + 13 = 27.

9 **(D) 5**
Let G be the number of grandchildren. Then 2 × G + 3 = 3 × G − 2. Solving the equation for G you get G = 5.

10 **(E)**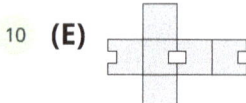

Notice that the cuts were made on two diagonally opposite edges of the cube.

4 Point Solutions

11 **(B) 12**

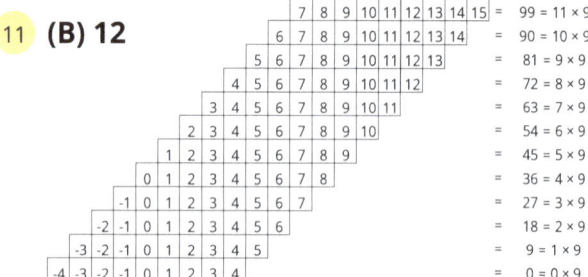

There are 12 such numbers: 0, 9, 18, 27, 36, 45, 54, 63, 72, 81, 90, and 99.

12 **(A) Sunday**
Drawing a calendar, we can find that the first Tuesday must have occurred on the 2nd (so that the Tuesdays are on the 2nd, 16th, and 30th). Thus the 21st is a Sunday.

13 (C) $125
Let *T* be the price of the tent.
60% of *T* + 40% of 40% of *T* + 30 = *T*
0.6*T* + 0.16*T* + 30 = *T*
Solving for *T* we get *T* = 125.

14 (D) 30
Let *G* be a number of green aliens, *O* be a number of orange aliens, and *B* be a number of blue aliens.
$2G + 3O + 5B = 250$, $B = G + 10$, and $O = G$.
Substituting, we have $2G + 3G + 5(G + 10) = 250$. Solving for *G* we get $G = 20$.
Finally, $B = G + 10 = 30$.

15 (B) 144
We get the least number of jumps if the kangaroo jumps 142 times the length of 7 ft, 1 time the length of 4 ft, and 1 time the length of 2 ft:
$142 \times 7 + 1 \times 4 + 1 \times 2 = 1000$.
Adding the jumps, we get: $142 + 1 + 1 = 144$.

16 (B) 18
The large square has a side equal to 3 small white squares. Thus, the length of the side of the large square and a small white square (or 4 small squares) is the same as three times the length of a side of a gray square. 4 sides of a small white square equal 24 ($3 \times 8 = 24$), so one side of a small square is 6. Hence, the large square has a side of length $3 \times 6 = 18$.

17 (B) 6
Let *x* be the number we are looking for. Then

$$x^2 = x + 500\%x$$
$$x^2 = x + 5x$$
$$x^2 - 6x = 0$$
$$x(x - 6) = 0$$
$$x = 6$$

Note that zero cannot be a solution because the number must be positive.

18 (C) 36°

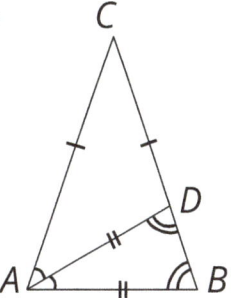

We know that $\angle ABD = \angle ADB$ and $\angle CAB = \angle CBA$. Because $\angle CAB$ has been bisected, $2 \frac{1}{2} \times \angle ABC = 180°$ and $\angle ABC = 72°$. Then because the angles of triangle *ABC* add up to 180°, we know that $\angle ACB = 180° - 72° - 72° = 36°$.

19 (A) 124
There is a pattern in the squares. 1st square: $1 \times 2 + 1 \times 2 = 4$, 2nd square: $2 \times 3 + 2 \times 3 = 12$, 3rd square: $3 \times 4 + 3 \times 4 = 24$, ... , 30th square: $30 \times 31 + 30 \times 31 = 1860$, 31st square: $31 \times 32 + 31 \times 32 = 1985$. Subtracting, we get $1984 - 1860 = 124$.

20 (B) 10
Let our original number be $\overline{ab2}$. Then $\overline{ab2} - 36 = \overline{2ab}$. We can easily find that $a = 2$ and $b = 6$. The sum of the digits of the number: $2 + 6 + 2 = 10$.

5 Point Solutions

21 (E) 12
Consider the following diagram, which indicates the number of ways that each tile can be reached. Notice that each tile is the sum of the number immediately above it and the number to its left, because she can only move down and to the right.

1	1	1	1	1
1	2	3	4	1
1	1	1	4	5
1	1	2	2	7
1	1	3	5	12

22 (B) 60

First, we know there are 5! = 120 ways to rearrange the cars. Car I and car II can be distributed uniformly randomly, so there is a 50% chance that I is ahead of II. Therefore there are 120 × 0.5 = 60 ways.

23 (E) 8

We want to keep the digits as large as possible to make the number of digits low. 2006 = 8 + 222 × 9; therefore the first digit of smallest natural number is 8. The number consists of one digit of 8 and 222 digits of 9.

24 (D) 3

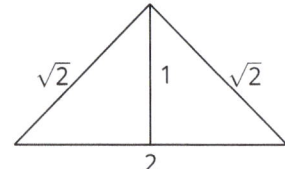

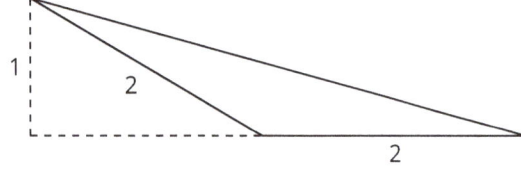

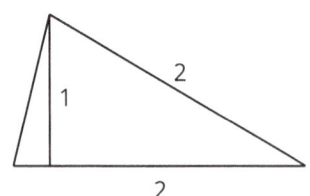

25 (B) 5

Let d be the distance, V be the speed, t be the time and k be the number by which the time would be shortened. Since $d = V \times t$ then $t = \frac{d}{V}$, and notice that $d = (V+3) \times \frac{t}{3}$. Substituting for t we have $d = (V+3) \times \frac{d}{3V}$. Solving for V we get $V = \frac{3}{2}$. The equation for the distance when Peter increases his speed by 6 m/s will be $d = (V+6) \times \frac{t}{k}$. Since the distance is equal we may look for k by solving the following equation by substituting:

$$(V+6) \times \frac{t}{k} = (V+3) \times \frac{t}{3}$$

$$\frac{15}{2} \times \frac{t}{k} = \frac{9}{2} \times \frac{t}{3}$$

$$k = 5$$

26 (C) can be divisible by 5.

Let $C = 2^5 \times 3 \times 5^2 \times 7^3$ and let $C = A \times B$ where $A = 2^5 \times 3 \times 5$ and $B = 5 \times 7^3$. Number A is divisible by 5 and number B is divisible by 5. Their sum $A + B$ is also divisible by 5.

27 (D) 54

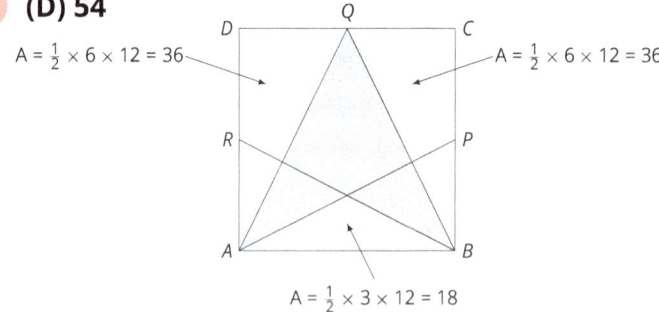

The area of the shaded region in in² is equal to 12 × 12 − (36 + 36 + 18) = 144 − 90 = 54.

28. (E) ANMAIKOLIRJ

M	I	S	S	I	S	S	I	P	P	I
K	I	L	I	M	A	N	J	A	R	O

P	S	I	S	I	M	I	S	S	P	I
1	2	3	4	5	6	7	8	9	10	11

The 1st and the 10th card can be A or R. This eliminates answers (A), (C), and (D). The 8th card cannot be J. This eliminates answer (B).

---

29. (C) 2005

Note that $1 \times 3 = 2^2 - 1$, $2 \times 4 = 3^2 - 1$. Thus the last 2004 terms of the first sum will all cancel to 1, so $1^2 + 2004 \times 1 = 2005$.

---

30. (E) None of the statements is true.

By finding a counterexample we prove the following:

(A) $10 \times 10 > 99$
($10 + 10 + 0.1 = 20.1$ and $10 \geq 10 \geq 0.1$)

(B) $20.055 \times 0.045 < 1$
($20.055 + 0.045 + 0.01 = 20.1$ and $20.005 \geq 0.045 \geq 0.01$)

(C) $15 \times 5 = 75$
($15 + 5 + 0.1 = 20.1$ and $15 \geq 5 \geq 0.1$)

(D) $18 \times 25/18 = 25$
($18 + 25/18 + 32/45 = 20.1$ and $18 \geq 25/18 \geq 32/45$)

Therefore (E) is the right answer.

2008

2008

3 Point Solutions

1. **(B) (1 + 2) × (2007 × 2008)**
Compare the following (1 + 2) > (1 × 2); (2007 × 2008) > (2007 + 2008); and (1 + 2) × (2007 × 2008) > (1 + 2) + (2007 + 2008).

2. **(C) 2**
There are 9 boys and 13 girls, which is a total of 22 students. Half of them have a cold, so 22 ÷ 2 = 11 have a cold. Since there are only 9 boys, at the very least 11 − 9 = 2 of the girls have a cold.

3. **(B) $\frac{1}{3}$**
Substitute 2 in the second expression to obtain
$$\frac{1}{1+2} = \frac{1}{3}$$

4. **(C) 5**
It takes 50 seconds to complete one full circle. Divide 50 by 10 to get 5.

5. **(C) 6**
We need to be able to make either 6 or 9 from the numbers already given. 6 = 4 + 2, and we can't make 9, so the missing number is part of the sum that makes 9. The other given number is 3, so the missing number is 9 − 3 = 6.

6. **(E) 1**
The first sum is 2 + 0 + 0 + 8 = 10, then 1 + 0 = 1.

7. **(B) 24 cm**
Since the two perimeters are equal to 16 cm, we know that the two sides of the triangle add up to 16 − 4 = 12 cm. Then, adding 12 + 4 + 4 + 4 = 24 cm.

8. **(B) 6**
Find the greatest common divisor of 24, 42, and 36. It is 6.

9. **(E) 14**
Adding the sides of the original cube and the new faces we have 6 + 8 = 14.

10. **(A) 52°**
First find the two angles on the bottom. 180° − 124° = 56°; 180° − 108° = 72°. Adding 72° and 56°, we have 128°. Then subtract 180° − 128° = 52°.

4 Point Solutions

11. **(C) 4**
Connecting all of the dots with segments we have 3 small squares and one bigger, skewed one in the center.

12. **(B) 5**
Dan has 9 × 2 = 18 cents and Ann has 8 × 5 cents. The sum of their money is 58, so to have the same amount they must each have 29 cents. To get that Dan must give Ann 2 coins (4 cents) and she must give Dan 3 coins (15 cents).

13. **(A) 10**
The digit of ones will be four times the digit of thousands again in the year 2018. So 10 is the minimum number of years when that situation will occur again.

14. **(C) 2**
The two pairs of numbers are (2 and 6) and (5 and 3).

15. (A) 1806
The answer has to be between 40 and 50 to fit in the answers provided. By trial and error we can try $x = 43$. Squaring 43 we get 1849. Subtract his age $1849 - 43 = 1806$.

16. (A) 5°
Let $x = \angle PBC$. Then $\angle PCB = 50° + x$. Solving the equation $x + 50° + x + 120° = 180°$ (the sum of angles in a triangle is 180°) gives $x = 5°$.

17. (C) 60 cm
Let the sides of the rectangles be a and b. The equation for the perimeter of each of the first two rectangles is $b + 2a = 40$. Similarly, the perimeter for each of the second two rectangles is $a + 2b = 50$. Setting up a system of equations
$$b + 2a = 40$$
$$a + 2b = 50$$
we get $3a + 3b = 90$, dividing by three we get $a + b = 30$. For the perimeter of the original rectangle we get $2a + 2b = 60$.

18. (A) 25
Start by drawing the segment $BC = 11$, then $CD = 14$, which gives $CA = 2$ and $DA = 12$. The distance $BD = 25$.

19. (D) 108 cm²
The base of the triangle is three times the radius of the circle, which equals 18 cm. The height of the triangle is twice the radius, measuring 12 cm. Using the area formula, we calculate the area as $1/2(12)(18) = 108$ cm².

20. (D) 3 and 5
Net #3 and net #5 will not form the cube because the placement of the triangles on the net will not result in a formation of a complete side. In net #3 one side would have two overlapping triangles, in net #5 one side would be missing one triangle.

5 Point Solutions

21. (D) 61
There are 125 unit cubes that make up the bigger cube. Each side is made up of 25 unit cubes. At most we can see three sides in a picture. Taking a picture we see 25 distinct unit cubes on the front wall, the right side we see 20 (since we already counted five of them on the front), on the top we see 16 (we have already counted five from the front and four from the right side). Adding $25 + 20 + 16 = 61$.

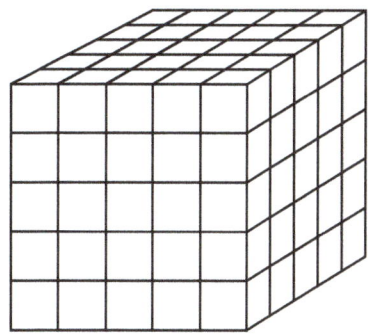

22. (C) 105
Using the examples we write $6 * 5 = (6 + 5 + 4 + 3 + 2 + 1) \times 5 = 21 \times 5 = 105$.

23. (C) 108°
Let $\angle BCD = x$. Then $\angle DCA$ also equals x. Since triangle ABC is isosceles we know that $\angle CBD = 2x$. Also, since triangle BCD is isosceles, $\angle BDC$ is equal $2x$. Solving the equation $2x + 2x + x = 180°$ (the sum of angles in a triangle equals 180°), we get $x = 36°$. Then $\angle BDC = 72°$. So $180° - 72° = 108°$.

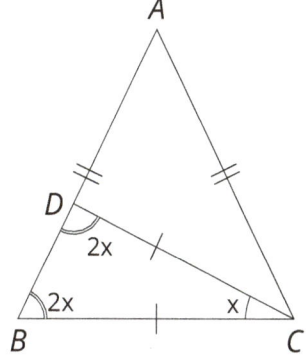

24. (D) 864
For a maximum value of *KAN*, we want to try the maximum value of *OO* = 99. To obtain the greatest *KAN* we use 8 for *K*. Then in order to get 99 we use 7 for *G*. To get 99, *A* is 6 and *R* needs to be by one bigger than *N*. Thus *R* is 5 and *N* is 4. For *KAN* we obtain 864.

25. (C) 5
Test each answer, making sure your fraction is less than 1/2 = 50%. A) 3/7 = 43%; B) 4/9 = 44%; C) 5/11 = 45.5%; D) 6/13 = 46% and E) 7/15= 47%. The smallest number that gives a percentage between 45% and 50% is 5/11, so the number of girls is 5.

26. (A) John
Since the boy tells the truth on two consecutive days, Thursday and Friday, the two possibilities are either John or Bob. If we assume that Bob is the answer, counting the answers we get that he answered Bob on Tuesday, which is a lie. Therefore, the answer is John.

27. (B) 40 minutes
Let *x* be the time during which car travels from *B* to *C*. Setting up a proportion comparing the truck and the car we get

$$\frac{90}{60} = \frac{60}{x}$$

and solving for *x* = 40 minutes.

28. (D) Set *B* has 2 times as many elements as set *A*.
Set *B* has 2 times as many elements as *A*. Set *A* consists of: (5511, 5151, 5115, 1155, 1515, and 1551); set *B* consists of (5311, 5131, 5113, 1153, 1513, 1531, 3115, 3151, 3511, 1351, 1315, and 1135).

29. (C) *BC* = *BD*
Angle *DBC* = 90 since it forms the same arc *CD* as angle *DAC* (*DAC* = 135° − 45° = 90°). Since angle *B* is 90°, we have a right isosceles triangle *CBD* where *CD* is the diameter of the circle so *BC* = *BD*.

30. (B) 12
The sum of any two even numbers will give an even answer, and the sum of any two odd numbers will give an even answer. The first sage must know that all the cards other than his are either all odd or all even. He must have pulled out all three even numbers to know that the second sage will have a pair of two odd numbers. So, the first sage must have the cards 2 + 4 + 6 = 12.

2010

2010

3 Point Solutions

1. **(C) 404**
   Adding the numbers gives 404. To make adding easier, we can group the numbers so the end digits add up to 10.

---

2. **(C) 2**
   An axis (plural: "axes") of symmetry is a line that divides a figure into two symmetrical parts in such a way that the figure on one side is the mirror image of the figure on the other side. In this image, we can find only two axes, as shown in the illustration.

---

3. **(D) 4**
   It is important to keep in mind that both the toy boxes and the larger cartons are cubes. In the picture shown, we initially have 8 cube boxes of toys. We can rearrange them to form a bigger cube.

   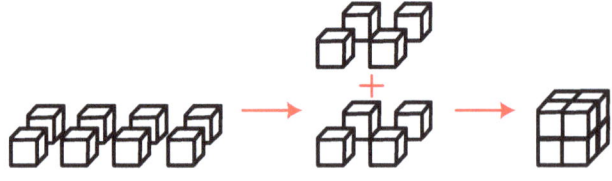

---

4. **(A) 6a + 8b**
   It is helpful to visualize the given figure as shown. Now that the length of each side is known, the perimeter of the figure can be derived as follows: $2(3a + 4b) = 6a + 8b$.

   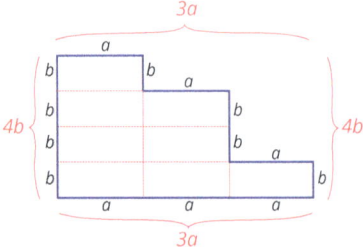

---

5. **(B) square**
   As shown in the figures, all the shapes listed except a square can be obtained by connecting some of the points of the hexagon.

   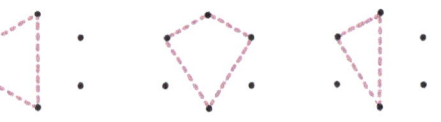

---

6. **(C) 2010**
   In this problem, the arithmetic mean for a nine-number sequence is 2006. So, the middle (5th) number in the sequence is 2006. If 2006 is the 5th number in the sequence given, it can be deduced that the sequence starts with 2002 and ends with 2010, which then is the largest number of the sequence.

---

7. **(E) 30**
   Grandmother must be able to split the cookies into 3, 5, or 6 equal amounts, so we are looking for numbers divisible by 3, 5, and 6. The fastest way to select some of the answers is to look for numbers divisible by 5. All numbers divisible by 5 end in 0 or in 5, so they are easy to spot. This approach gives you two answers to choose from: (B) 15 and (E) 30. Only (E) 30 is also divisible by 3, 5 and 6.

---

8. **(B) 3**
   Only 3 blocks need to be moved. Look at the following sequence:

9 (B) 6

There is a square that has only blue tiles, one that has one green tile, two that have two green tiles, one that has three green tiles, and one that has only green tiles. All the possible non-identical squares are illustrated.

10 (C) 100

It is helpful to look at the equation this way: $(2 − 1) + (4 − 3) + (6 − 5) + ... + (100 − 99) = 1 + 1 + 1 + ... + 1 = 100$. The subtraction of any odd number from the following even number will always be 1. Because there are 100 subtractions, each resulting in 1, the sum of all those would be 100.

4 Point Solutions

11 (C) 19

Working in reverse, assume that each cut removes one piece of wood. After undoing 53 cuts, and so 53 removing pieces of wood from the final 72, 19 pieces are left. The key to this problem is to realize that each cut produces one piece of wood.

12 (D) 25

One way is to quickly add the 3 largest different single digit numbers: $7 + 8 + 9 = 24$. Answer (D) 25 cannot be obtained by the sum of three different single digit numbers. Now, check for a possible combination for the next smallest number 23: $6 + 8 + 9 = 23$. Thus, the answer is (D) 25.

Another way is to try to find a combination of 3 different single digit numbers that add up to each of the given answers, starting with the smallest.
10 = 4 + 5 + 1 (one possibility)
15 = 6 + 7 + 2
23 = 8 + 9 + 6
25 = N/A

13 (D) 45 min

When joining three chain segments together, there are two joining points. If it takes the blacksmith 18 minutes to complete two joining operations, a single joining operation takes 9 minutes. If the blacksmith has to join 6 segments of chain, he would have to perform 5 joining operations, 9 minutes each. Thus, the total time needed is: $5 × 9$ min $= 45$ min.

14 (B) 55°

In the triangle ACD, $180° = ∠ADC + 50° + 65°$. Therefore $∠ADC = 65°$. Since $∠ADC = ∠ACD = 65°$, it follows that $|AD| = |AC|$. Then because $|AD| = |BC|$ is given, we have that $|AD| = |AC| = |BC|$. Therefore in triangle ABC, $∠ABC = ∠BAC$.

Then in triangle ABC, we have $2 ∠ABC + 70° = 180°$. Therefore $∠ABC = 55°$.

15 (C) 19

Let w be the number of white blocks, b blue, and r red. Then we have that $w + b + r = 50$, $w = 11b$, $w > r$, and $r > b$. From the multiplication factor for $w = 11b$, we can assume that the value for b is fairly small. If we assume that $b = 2$, it would follow that $w = 11 × 2 = 22$. From the first given equation, we can derive that $22 + 2 + r = 50$, $r = 26$. Since w has to be greater than r, the assumed value for b does not satisfy all conditions.

Let us now assume that $b = 3$, then $w = 11 × 3 = 33$. If $33 + 3 + r = 50$, $r = 14$. In this case, r is smaller than w and greater than b. Thus, the values for w, b, and r satisfy all conditions. Now, to calculate the difference between w and r, subtract their values: $w − r = 33 − 14 = 19$. Thus the correct answer is (C) 19.

16 (A)

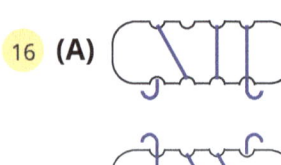
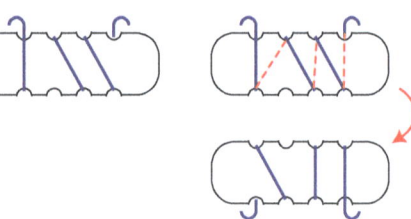

Look at the figure that has been given, and draw dashed lines representing the wire on the back side. Then, flip the board and find the corresponding figure in the answers, (A).

---

17 (A) 1

We know that area $ABCD = BC \times CD = 6 \times 10 = 60$. We also know that area $XYSR = XS \times SR = 6XS$ since $PQRS$ is a square. It is given that area $ABCD = 2 \times XYSR \Rightarrow 60 = 2 \times (6XS)$. Therefore, $XS = 5$. We have that $PX = PS - XS = 6 - 5$, and thus $PX = 1$.

---

18 (B) 4

A minimum of 4 straight lines is needed to divide a plane into 5 regions. Note that lines and planes extend infinitely.

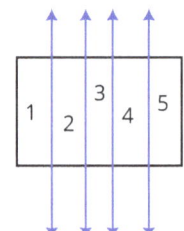
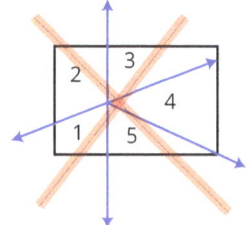

---

19 (E) e

The relationship between a, b, c, d, and e is given by the equation:
$a - 1 = b + 2 = c - 3 = d + 4 = e - 5$.
One way to look at it is to first separate b and d. Since both are defined by the addition of another number, we can assume that b and d are the two smallest numbers in the group. Now, we are left with: $a - 1 = c - 3 = e - 5$. The largest among a, c, and e would be the one that needs the largest number subtracted in order to be equated to the rest. Since $1 < 3 < 5$, the correct answer is (E) e.

---

20 (B) $\frac{1}{4}$

The area of a circle is $A = \pi r^2$. We know that the whole figure is half a circle with the biggest given radius, 8 cm. Thus, the area of the whole figure is

$A = \pi \times 8^2 \div 2 = 32\pi$

Similarly, the shaded area is half of a circle with radius of 4 cm. Thus the area of the shaded circle = $\pi \times 4^2 \div 2 = 8\pi$.

Thus the ratio of shaded area to total area is
$\frac{8\pi}{32\pi} = \frac{1}{4}$

---

5 Point Solutions

21 (C) 64

The sum of two whole consecutive numbers is always an odd number. The number 103 can be represented as 51 + 52. The rest of the given numbers have to be the sum of at least three numbers of roughly the same magnitude. This means that you can try to divide the number by 3 or 4 and so on to get the middle number of the sum we are looking for. We find that $14 = 2 + 3 + 4 + 5$, $24 = 7 + 8 + 9$, $2010 = 400 + 401 + 402 + 403 + 404$. Only 64 is not equal to the sum of at least two positive consecutive whole numbers.

---

22 (C) 16

We are given four types of poultry: turkey (t), goose (g), chicken (c), and hen (h).
We have the following relationships:
$t = 5c$, $g + 2h = 3c$, $g = 4h$. Since the problem is looking for how many hens are needed to trade for one turkey, one goose, and one chicken, we have to define t, g, and c in terms of h. Considering $g + 2h = 3c$, we can plug in $g = 4h$ to obtain $6h = 3c \Rightarrow c = 2h$. We are given $t = 5c$, so plugging in the result we just obtained, we get $t = 10h$. Therefore $t + g + c = 10h + 4h + 2h = 16h$.

23

(E) the point of intersection of the height of triangle *ABC* at angle *A* with side *BC*.

The following illustrations help visualize the geometry described by the problem. The conditions given by the problem are applicable to all types of triangles. The examples here include both an obtuse and an acute triangle. As shown, if *P* is the second intersecting point of the two circles, then *AP* is the height of the triangle *ABC* at angle *A* with side *BC*.

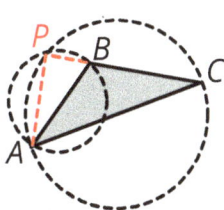
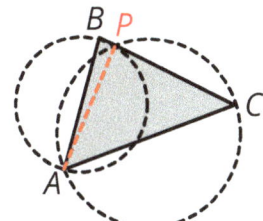

24

(B) 5

One way to look at this problem is to create an equation that can be checked with the possible answers given. If x is the number of cards labeled 4, and y is the number of cards labeled 5, then $4x + 5y = S$, the sum. We have that

$(5 - 1)x + 5y = S$
$5x - x + 5y = S$
$5(x + y) - x = S$
$90 - x = S$ since $x + y = 18$

is the number of cards given.

We know that S is divisible by 17. What value for x is needed to make the sum divisible by 17? Let us try some of the answers: if $x = 4$, then $86 \div 17$ does not give a whole number. However, $x = 5$ gives $85 \div 17 = 5$, which is correct.

25

(C) 36 cm

Draw a right triangle with a hypotenuse connecting the centers of these circles. AC is equal to the sum of the two radii; $AC = 17 + 9 = 26$. AB can be found by using the given length of the rectangle, 50 cm. As the illustration shows: $AB + 17 + 9 = 50$ cm. $AB = 50 - 17 - 9 = 24$ cm. BC can be found by the Pythagorean Theorem: $AB^2 + BC^2 = AC^2 \Rightarrow 24^2 + BC^2 = 26^2 \Rightarrow BC = 10$. To find the side of the rectangle, $17 + 10 + 9 = 36$ cm.

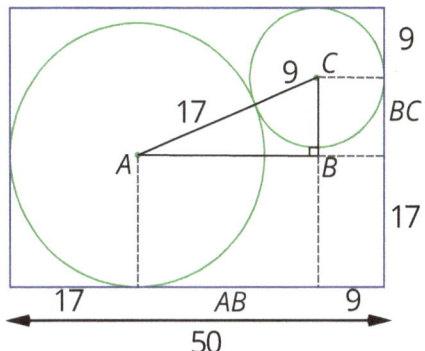

26

(C) 4 people, 2 liars

The key to this problem is to remember that knights *always* tell the truth, and that liars *always* tell lies. Each of the given statements of the townspeople has two parts. That means that if one part of the statement is true, the second is true as well.
- The statement, "Each of us is a liar," is untrue, because then the statement itself would be a lie, so there are more than 3 people in the room.
- The statement, "Not everyone is a liar," is true. So, we can assume that the first part of the statement is true as well: "There are at most 4 people in the room."
- The statement, "There are 5 people in the room," is untrue, because we saw in the second statement that there are 4 people in the room.

In conclusion, there are 4 people in the room. The persons who had given the first and third statements are liars, so 2 of the 4 people in the room are liars. (There could also be 3 liars among the 4 people, but that is not one of the answers listed.) The correct answer is (C).

27 (B) 8

In the illustration, a cube is constructed from 27 smaller cubes. In the problem, each vertex of the cube must be shared by cubes of different color. Some vertices, like the ones on the surface, are shared only by 2 or 4 small cubes. To determine how many different colors are needed, let us take the worst case scenario. The small cube in the middle would have vertices that are shared with the largest number of other small cubes, so that there are 8 cubes that share a vertex.

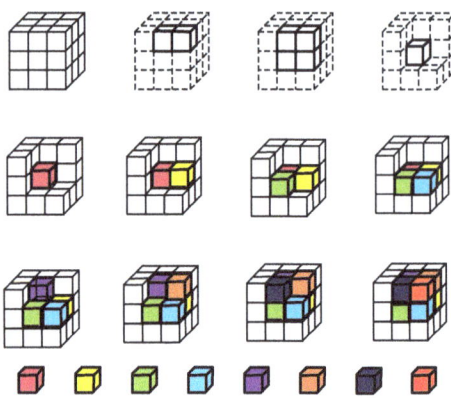

Eight different colors are needed for the figure to be constructed in a way so that cubes sharing a vertex are of different colors.

---

28 (A) 11

One way to look at this problem is to divide the triangle ABC into three areas, A_1, A_2, and A_3. As shown in the illustration, each of those areas is a smaller triangle that is half of a parallelogram.

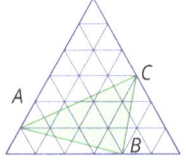 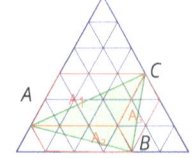

To find the area of triangle ABC, we need the sum of A_1, A_2, and A_3. $A = A_1 + A_2 + A_3$.
To find A_1, simply count the number of equilateral triangles in the parallelogram. Since A_1 is half the area of the parallelogram, divide the number of triangles by 2. Do the same for triangles A_2 and A_3. $A_1 = 12 \div 2 = 6$, $A_2 = 6 \div 2 = 3$, $A_3 = 4 \div 2 = 2$. Thus $A = 6 + 3 + 2 = 11$.

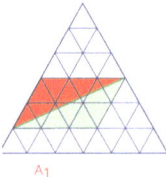 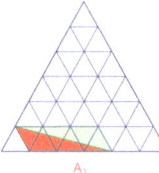 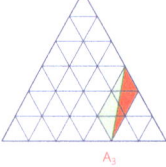

29 (D) $\frac{6}{7}$

Let's check each answer.
(A) $\frac{7}{8}$ is possible. Choose $y = 7$, $x = 8$.
LCM(24, 8) = 24 < LCM(24, 7) = 24 × 7
(B) $\frac{8}{7}$ is possible. Choose $y = 8 \times 24$, $x = 7 \times 24$.
LCM(24, 7 × 24) = 7 × 24 < LCM(24, 8 × 24) = 8 × 24
(C) $\frac{2}{3}$ is possible. Choose $y = 8 \times 2$, $x = 8 \times 3$.
LCM(24, 8 × 3) = 24 < LCM(24, 8 × 2) = 48
(D) $\frac{6}{7}$ is not possible.
Suppose there exist y and x such that $\frac{y}{x} = \frac{6}{7}$. Then there exists an integer k such that $y = 6k$ and $x = 7k$.
LCM(24, x) = LCM(24, 7k) = 7 LCM(24, k). This is because there are no factors of 7 in 24.
LCM(24, y) = LCM(24, 6k) ≤ 6 LCM(24, k). This is because LCM(24, 6k) = 6 LCM (4, k) and LCM(4, k) ≤ LCM(24, k) because 4 is a factor of 24. From this we see that LCM(24, x) > LCM(24, y) for all x and y such that $\frac{y}{x} = \frac{6}{7}$.
(E) $\frac{7}{6}$ is possible. Choose $y = 7$, $x = 6$.
LCM(24, 6) = 24 < LCM(24, 7) = 24 × 7

---

30 (D) 35

A good strategy in this problem is to make an equation, which then is used to try out some of the answers. All 5 students brought different numbers of pieces of candy. It is safe to say that the smallest possible sum of all the pieces of candy would be the sum of 5 consecutive numbers because this minimizes the sum while also ensuring that all students have different numbers. We can express consecutive numbers in the following equation:
$a + (a + 1) + (a + 2) + (a + 3) + (a + 4) = S$.
Then $5a + 10 = S$. Let us consider small values of a. If $a = 1$, then the students bringing 4 and 5 will bring more than the students bringing 1, 2, and 3. Trying increasing values of a, we find that when $a = 5$, $5 + 6 + 7 > 8 + 9$, and thus the minimum pieces of candy brought is 35.

2012

2012

3 Point Solutions

1 **(B) 2 dollars**
Since four chocolate bars are three more than one chocolate bar, the price difference of 6 dollars corresponds to three chocolate bars.

2 **(D) 9.999**
11.11 − 1.11 = 10.00, so 11.11 − 1.111 = 10.00 − 0.001 = 9.999

3 **(A) 45**
Every 15 minutes, the watch will move 90 degrees, or go to a new direction. In 15 minutes it will go from northeast to southeast, in another 15 it will go from southeast to southwest, and in a third interval of 15 minutes it will go from southwest to northwest. So, in all, it will take 45 minutes for the minute hand of the watch to point northwest.

4 **(E)**
The O can only be cut into two different pieces. The F, S, and H can each be cut so that they fall apart into at most four pieces, and the M can be cut so that it falls apart into five pieces.

5 **(C) 29**
Each time a head is chopped off, one is removed and five are added, so that the total change is the addition of four new heads. If five heads are there to begin with, and 4 × 6 = 24 new heads are added in this way, then 29 heads will be the final number.

6 **(E) (8 + 8 − 8) ÷ 8**
In the expression corresponding to (E), we have the form: $(X + X - X) \div X = X \div X = 1$. This means that for any number given, the expression will be equal to 1.

7 **(C) 700 m**
If Ann starts at A and ends at B, then she must only travel along one of the two paths touching A, and only one of the two paths touching B, or else she will either not make it to her destination, or have to retrace her steps. With two paths eliminated in this way, seven remain for her to cross in a variety of different ways. Since each path is 100 m long, the total length of a longest possible route is 100 × 7 = 700 m.

8 **(D) 4**
There are four different ways. One could either choose the top right vertex of each triangle, the bottom right vertex of each triangle, the top right vertex of the left-hand triangle and the left vertex of the right-hand triangle, or the bottom right vertex of the left-hand triangle and the left vertex of the right-hand triangle. Any other combination of two vertices will either be on the same triangle, or determine a line which crosses through one or both triangles.

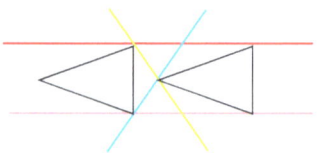

9 **(D)**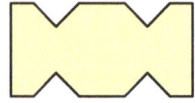

Figure D is the only figure which would require four straight cuts to make, given that the piece of paper has been folded over just once.

    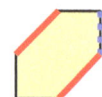

10 (D)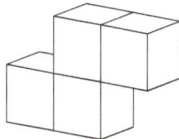

We are shown two white cubes, and must determine the locations of two more to solve this problem. Since the black piece must have four cubes, and only three are shown, the fourth one must be below the right-hand white cube shown. This means that one of the white cubes must be below the left-hand white cube shown, as this is the only cube touching either of the white cubes shown that is not necessarily taken by another cube. Since we are also shown four dark gray cubes, the last white cube must be below the gray cube which only has its top face visible, since this is the only space not accounted for by this point.

4 Point Solutions

11 (C) 3825
To minimize the sum of the two numbers, the 1 and 2 digits need to be used in the thousands places of the two numbers being summed, the 3 and 4 in the hundreds places, the 5 and 6 in the tens, and the 7 and 8 are left over for the ones places. Since addition is commutative, the specific combinations of numbers do not matter. Any numbers in the form $\overline{abcd}$ work, where $a = 1$ or 2, $b = 3$ or 4, $c = 5$ or 6, and $d = 7$ or 8. For example, $1357 + 2468 = 3825$.

12 (C) 10 m²
The diagram in this problem is misleading and purposely not drawn to scale. Since we are given that the area of the strawberry bed has been reduced by 15 square meters by having one of the sides shortened by 3 meters, we know that the length of the other side must be $15 \div 3 = 5$ meters. This length is also the length of one of the sides of the square which is the pea bed in this current year. Since the pea bed has been lengthened by three meters this year, in the previous year it measured 5 meters by $5 - 3 = 2$ meters, which would have made its area equal to $5 \times 2 = 10$ square meters.

13 (B) 60
If we call the three inserted numbers a, b, and c, respectively, then we know the following facts about them: $10 + a + b = 100$, so $a + b = 100 - 10 = 90$, $a + b + c = 200$, and $b + c + 130 = 300$, so $b + c = 300 - 130 = 170$. To solve the problem, we need to combine the various equations. For example, since $a + b = 90$, and $a + b + c = 200$, we know that c alone must be equal to 110, since when you subtract $a + b$ from $a + b + c$, you also subtract 90 from the 200 that it is equal to. If c is equal to 110, then $b + c = 170$ means that $b = 170 - 110 = 60$.

14 (C) 51
To solve this problem, one must recognize that the supplement of 100° is 80°, and that this is the measure of one of the angles of the triangle which has one angle equal to 58°. Therefore, the third angle in this triangle must measure $180° - 80° - 58° = 42°$. Because vertical angles have the same measure, the triangle containing angle x will also have one angle measuring 42°. Since we can also see that it will contain 87°, the supplement to 93°, we can figure out that x is $180° - 87° - 42° = 51°$.

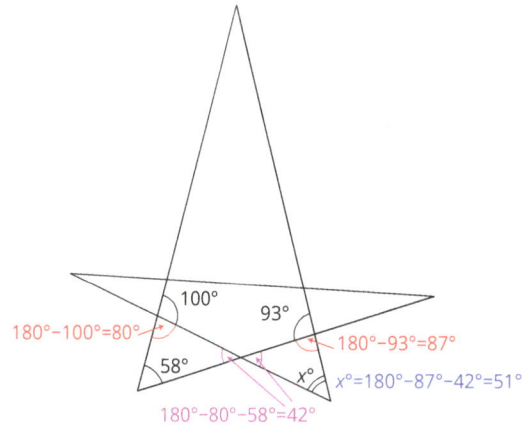

15 (C) 7
Since it is true that the number 7 is prime, odd, and divisible by 7, the only phrase which can be on the back of the card which has 7 on one side is "greater than 100."

16 (D) 1.5 cm

Since the perimeters are the same, we can look at one of the long edges of the remaining hexagon and compare it to two of the three edges of one of the small triangles, since there are 3 such long edges, which must be equal to 6 small edges of the triangles. (The other, smaller edges of the hexagon cancel out one edge on each of the small triangles, since they are the same edges used in both figures.) Since two small edges of one white triangle are equal to one big gray edge of the hexagon, we know that the original triangle had edges equal to 2 white edges plus a gray edge, or also equal to 2 white edges plus a gray edge which is together 4 white edges, which was equal to 6 cm. Therefore, one white edge, which is the side length of a small triangle, is $6 \div 4 = 1.5$ cm.

---

17 (C) 6

The largest number of mice is 6. To see this, note that three of the mice could have stolen 5, 7, and 9 pieces of cheese without one having stolen twice the number as any other. Two mice cannot have stolen both 3 and 6 pieces of cheese, but one might have stolen 3 (or perhaps 6). Considering 1, 2, 4, and 8 pieces of cheese, if any three of these numbers of pieces of cheese were stolen, then one mouse would have stolen twice as many pieces as another. However, two of these numbers of pieces of cheese could have been stolen (1 and 4, 1 and 8, or 2 and 8). This gives a maximum of six different numbers of pieces of cheese that could have been stolen.

---

18 (E) 300 m

Ann has a total speed of $6 + 4 = 10$ km/hour, which is two-and-a-half times faster than Bill, who only has a speed of 4 km/hour. This means that Bill only covers 40 percent of the distance that Ann does in a given period of time. When Ann has covered 500 meters, Bill has only covered $500 \times (40/100) = 200$ meters, leaving him $500 - 200 = 300$ meters behind Ann.

---

19 (D) 112

If a square's perimeter is doubled, then each of the side lengths must be doubled. To achieve the maximum final perimeter, the two doublings from lying must come first, and the two true statements must follow, to shorten each side by 4 cm. So, after quadrupling and then shortening each side by 4 cm, we have a final length of $8 \times 4 - 4 = 28$ cm. This means that the square, which has 4 sides, has a total perimeter of $28 \times 4 = 112$ cm.

---

20 (B) 1 and 6

If one views the cube from directly above, with the top of the page being "north" and the bottom "south," then the first two flips, from 1 to 2, and from 2 to 3, flip the cube upside-down. Flipping it from 3 to 4 then turns it on its side so that the face which was originally pointing vertically up is flipped from pointing vertically down to pointing "south." The flip from 4 to 5 simply rotates this face, so that when it is flipped from position 5 to position 6, it is pointing vertically up again. Therefore, if the same face was pointing vertically up in positions 1 and 6, then the same face, which is opposite of the one pointing vertically up, must have been occupying both of these positions.

5 Point Solutions

21 (E) 50 cm

Let h be the height of the smallest cube, so that the five cubes have heights h, $h + 2$, $h + 4$, $h + 6$, and $h + 8$. From the conditions of the problem, we must have $h + h + 2 = h + 8$, so that $h = 6$. Thus, the tower made from all five cubes is $6 + 8 + 10 + 12 + 14 = 50$ cm tall.

---

22 (D) 3:16

Since the triangles *ABC* and *ADC* are reflections of each other over the line *AC*, they are equal in area, and since the triangles *DMC* and *AMC* both have base length equal to $AD \div 2$ and height equal to *DC*, they are also equal in area. This means that the fraction of space which triangle *CMN* occupies in triangle *AMC* is one-fourth of the fraction of the space which it occupies in the entire square, since triangle *AMC* is a half of a half of the entire square. Since *MN* is perpendicular to *AC*, we know that *N* must be a quarter of the way to *C*, since if *M* were at *D*, *N* would be at the center of the square, and *M* is only halfway to *D*, so *N* must be only halfway to the center from *A*. Thus, if we rotate triangles *AMN* and *CMN* 45 degrees clockwise and compare them, we see that they have the same height, and that *CMN* has a base length three times greater than *AMN*. This means that triangle *CMN* takes up three quarters of the area of triangle *AMC*. Therefore, it takes up (3/4) ÷ 4 = 3/16 of the area of the entire square.

23. (B) 24
If we say that the number of men is equal to *M* and the number of women is equal to *W*, then we know that *M* + *W* is less than or equal to 50. We also know that 3*M*/4 = 4*W*/5, or that *M* = 16*W*/15. This means that for every 15 women, there are 16 men. Since people cannot be paired up in fractions, there must be some multiple of 31 people at the dance. 31 is therefore the number of people at the dance, since 62, 93, or any other multiple of 31 is greater than 50. This means that there are 16 men and 15 women. If there are 3/4 × 16 = 12 men dancing, then there must also be 12 women dancing. This means that there is a total of 24 people dancing.

24. (D) 6 and 8
Begin by considering the number 1. It must be the case that 3 and 4 are on either side of 1 given the conditions of the problem — it doesn't matter in which order, so assume that reading clockwise we have 4, 1, 3. Now 2 must be next to both 4 and 5, so this gives us the arrangement 5, 2, 4, 1, 3. Proceeding in this fashion, we see that the only arrangement of numbers which works (starting at 1 and going either direction through the other 11) is: 1, 3, 6, 8, 11, 9, 12, 10, 7, 5, 2, 4.

25. (D) 1993
The numbers which have the described property are numbers whose middle digit is both the first and second digit of two-digit numbers which are perfect squares. All of the two-digit perfect squares are: 16, 25, 36, 49, 64, and 81. Therefore, the only three-digit numbers with the described properties are: 164, 364, 649, and 816. The sum of these four numbers is 1993.

26. (E) 23
If the first story starts on the first page and has an even number of pages, then the second story will also start on an odd-numbered page, and so on for each next story, as long as the one before it had an even number of pages. In this way, the first 16 stories can each start on an odd-numbered page, by having the first 15 stories being even numbers of pages long, and the 16th starting and ending on an odd-numbered page. After this, the stories will alternate beginning on even- and odd-numbered pages, meaning that stories number 17, 19, 21, 23, 25, 27, and 29 will NOT begin on odd-numbered pages. Since 7 out of 30 stories do not begin on odd-numbered pages, the remaining 23 can begin on odd-numbered pages in the way just described.

27. (B) 4
If we denote the first position as zero degrees of rotation, then the first few positions will be: 0°, 3°, 3 + 9 = 12°, 3 + 9 + 27 = 39°, 3 + 9 + 27 + 81 = 120°, 3 + 9 + 27 + 81 + 243 = 363°, and so on. If two positions are off by any multiple of 120°, such as positions 1 and 5, or 2 and 6, then they are the same. The first 4 positions are considered unique, because none of the positions are greater than 60, preventing them from being integer multiples of 60 (zero does not count, as it is the first position and is considered unique). After these 4 initially unique positions, there are no more unique ones, since adding any higher multiple of 3 degrees is just like adding one of the smaller numbers of degrees. For example, 243 is just 240 + 3, and since 240 is 4 × 60, the 240 does not count, and this movement is just like adding 3°. This means that the rotations end up forming a pattern, where the actual rotation seen (in degrees) is +3, +9, +27, +81 (or -39), +3, +9, +27, +81 (or −39), +3, +9, +27, +81 (or −39), +3, +9, +27, +81 (or −39), . . ., and so on.

28. (C) 72 m

If the rope was folded three times, then its length was decreased by a factor of 8, into a coil of length x, which was then cut at some spot which was a distance of c away from the end with the loose ends of the rope and a distance of $x - c$ away from the other. Because of the coiling, the lengths of rope which could have resulted from this are c, $2c$, and $2(x - c)$. By considering the possible combinations of two of these four numbers for various original rope lengths, we can rule out the impossible length. For case (A), an original rope length of 52 would lead to a coil of length 6.5 m. This means that if the rope was cut 2 meters from one end, ($c = 2$ and $x - c = 4.5$), it would result in pieces of length 2 m, 4 m, and 9 m, which is allowed, since both 4 m and 9 m lengths of rope appear. For case (B), a 68 m rope would make a coil of length 8.5 m. This could be cut 4 meters away from one side so that $c = 4$ and $x - c = 4.5$. This would give rope lengths of 4 m, 8 m, and 9 m. This situation is also allowed. For case (C), the coil would have to be of length 9 m, for which there is no c which gives rope strands of the appropriate lengths. For case (D), the coil would be 11 m long, and a situation where $c = 9$ and $x - c = 2$ would lead to rope lengths of 4 m, 9 m, and 18 m, which is allowed.

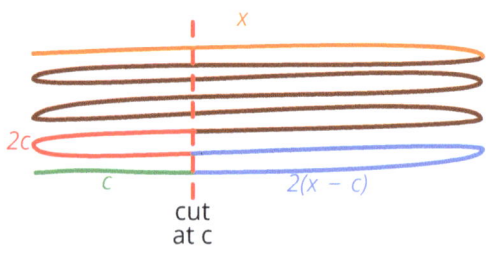

29. (C) 13

If one takes the sum of the perimeters of the three quadrilaterals (25 cm) and adds to it the perimeter of the four triangles (20 cm) and then subtracts the perimeter of the whole triangle (19 cm), then one is left with double the sum of the lengths of the three straight line segments, which is $20 + 25 - 19 = 26$ cm. Therefore, the sum of the lengths of the three straight line segments is 13 cm.

30. (A) 16

First, look at the product of the sixteen numbers in the four 2 × 2 squares. This product ($2^4 = 16$) uses the corners once each, the edge squares twice each, and the center square four times. Since the product of all the numbers in the square is 1, we can ignore one occurence of each square in each row, and get that the product of the four edge squares with the center square three times is also 16. But the product of the middle column and the middle row is also 1, so we can ignore the edge squares and two occurences of the central square, and see that the center square must be 16.

2014

2014

3 Point Solutions

1 **(D) March 21st**
The latest possible date of the competition is March 21st. Note that in order to have the latest possible 3rd Thursday the first day of the month should be Friday.

S	M	T	W	T	F	S
					1	2
3	4	5	6	7	8	9
10	11	12	13	14	15	16
17	18	19	20	21	22	23
24	25	26	27	28	29	30
31						

2 **(D) 4**
Labeling the three smallest regions A, B, and C, we have the rectangles: A, A and B, B, B and C. Thus there are 4 quadrilaterals.

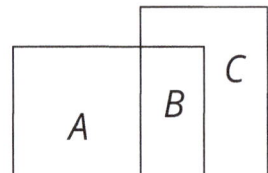

3 **(A) 0**

$$\frac{2014 \cdot 2014}{2014} - 2014 = 2014 - 2014 = 0$$

4 **(B) 5**
Notice that segment MN divides the rectangle in two equal parts and four congruent triangles. Thus the shaded part is $\frac{1}{2}$ of the rectangle, so MBND has area 5.

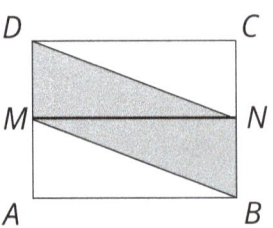

5 **(E) 35**
The following pairs of numbers have a product of 36: (1, 36), (2, 18), (3, 12), (4, 9), and (6, 6). The only pair of numbers whose sum is 37 is (1, 36). The difference of numbers 36 and 1 is 35.

6 **(E) 6**
The bird uses 4 small triangles, 1 big triangle, and 2 squares. Note that each small triangle is 1/8 of the original square, each big triangle is 1/2 of the original square, and each square is 1/4 of the original square. Thus the total area is 2 + 2 + 2 = 6.

7 **(B) 8 liters**
Note that 2 liters of a cleaner is equivalent to a quarter of a bucket. The capacity of the bucket is four times 2 liters, which is 8 liters.

8 **(E) 20**
In order to make a cube with a length of 3 units for each edge, a total of 3 × 3 × 3 = 27 little cubes are needed. In the picture, there are 7 little cubes used. The number of little cubes needed can be calculated as: 27 − 7 = 20.

9 **(B) 55 × 666**
Notice that all products are the results of a 2 digit number multiplied by a 3 digit number. Calculating the products of 4 and 7 (28), 5 and 6 (30), 7 and 4 (28), 8 and 3 (24) and 9 and 2 (18), it can be observed that the product of 5 and 6 is the largest. In the same way, 55 × 666 gives the largest result.

10 **(D) 7**
In order to take off the largest number of white beads and only five gray beads, take 6 beads from the left side and 6 from the right side, giving 7 white beads.

4 Point Solutions

11 (E) 10
Let the quarter have x weeks. Then Jack has $2x$ lessons and Hannah has $x/2$ lessons. Then $2x - x/2 = (3/2)x = 15$, and thus $x = 10$.

12 (B) $\frac{9}{2}$ cm²
Notice there are four overlapping parts of the circles, each with an area of $\frac{1}{8}$ cm². The region covered by the circles is $5 \cdot 1 - 4 \cdot \frac{1}{8} = \frac{9}{2}$ cm².

13 (C) 4
Number 100 has to be expressed as a sum of three numbers, each of which is a power of 2. The following sum shows this possibility: $2^2 + 2^5 + 2^6 = 4 + 32 + 64 = 100$. The granddaughter is 4 years old.

14 (E) 32 cm²
Consider the bottom length of the square; this length can be made by using 3 rectangles. Thus it has side length $24 \div 3 = 8$ cm. Similarly, by looking at the length of the right side of the square, we find $(24 - 2 \cdot 8) \div 2 = 4$ cm. So, the area of a rectangle is $8 \cdot 4 = 32$ cm².

15 (E) It will never happen.
In a heptagon (seven-sided figure), 3 places clockwise is the same as 4 places counter-clockwise. The heart and the arrow will always be 2 spaces apart after any of the described moves.

16 (C) 60°
Letting $\angle HBA = \beta$, we have that $2\alpha + \beta + 90° = 180°$. Using ABO, we have $\alpha + \beta + 4\beta = 180°$. Solving this system yields $\alpha = 30°$ and $\beta = 30°$. Thus $\angle CAB = 2\alpha = 60°$.

17 (B) 7:46
Optimally, bathroom 1 uses times of 21, 8, and 17 (46 minutes). Bathroom 2 uses times of 22, 10, and 12 (44 minutes). Thus they would finish at 7:46.

18 (E) 5 cm, 1 cm, 5 cm
By drawing the bisectors, we create isosceles right triangles with legs of length 6. Because the length is 11, then the long side will be split into lengths of 5, 1, and 5.

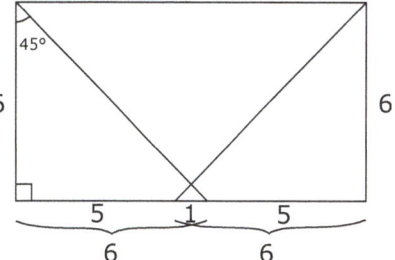

19 (D) 150
From, "if there had been 50 fewer coins, then each person would have received 5 coins less," we calculate there were 10 pirates since 50 divided by 5 is 10. From, "if there had been four pirates less, then each person would have received 10 more coins," we assume that 4 pirates received $10 \times 6 = 60$ coins. If 4 pirates received 60 coins then 10 pirates received x coins.
Solving $\frac{4}{10} = \frac{60}{x}$, we get $x = 150$.

20 (A) 75%
The equation that represents the problem is: $\frac{a+b}{2} = 0.7a$. Solving for a, we get: $a = \frac{5}{2}b$. We can express $\frac{a+b}{2}$ in terms of b as follows:

$$a = \frac{5}{2}b$$
$$a + b = \frac{5}{2}b + b$$
$$a + b = \frac{5}{2}b + \frac{2}{2}b$$
$$a + b = \frac{7}{2}b$$
$$\frac{a+b}{2} = \frac{\frac{7}{2}b}{2} = \frac{7}{4}b$$

Notice that 7/4 is equal to 1.75, which is 175%. The average is greater than the other number (b) by 75%.

5 Point Solutions

21 **(E) 27**

The way to have "neighbors" of 9 equal to 15 in the given cells is to place 9 between 3 and 4 and have the third "neighbor" equal to 8 (8 + 3 + 4 = 15). This means that 8 is in the middle cell. The remaining numbers: 5, 6 and 7 can be placed in any order into the missing cells and the sum of the "neighbors" of 8 will be: 5 + 6 + 7 + 9 = 27.

---

22 **(D) D**

From the information:
B + E = 800 and B + C = 900, we conclude that E < C.
From the information:
B + D = 1200 and B + E = 800, we conclude that E < D.
From the information:
C + E = 2100 and A + E = 700, we conclude that A < C.
From the information:
B + D = 1200 and B + C = 900, we conclude that C < D.
From the information:
C + E = 2100 and B + E = 800, we conclude that B < C.
Summarizing:
E < C, A < C, B < C, E < C, and C < D, we conclude that the heaviest weight is weight D.

---

23 **(B) 45**

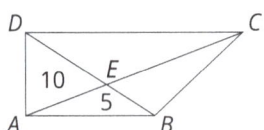

Notice that ABC has the same area as ABD, as both have the same base and height. Now call the center intersection E. ABD is equal to 15 and the area of BCE is 10. Also having in mind that two diagonals of any trapezoid divide it into four proportional triangles, the following proportion is true for the given triangles in the trapezoid:

$\frac{AED}{ECD} = \frac{ABE}{BCE}$ and $\frac{10}{ECD} = \frac{5}{10}$

Solving, we get ECD = 20. Thus ABCD has an area of 10 + 5 + 10 + 20 = 45.

---

24 **(D) 56**

Let's mark by x a number of questions worth 1 point, which also means questions that were solved by both girls. Let's mark by y a number of questions worth 4 points. The following system of equations takes place: $1x + 4y = 312$ and $x + y = 120$. Solving for x and y we obtain $x = 56$ and $y = 64$. There were 56 questions solved by both girls.

---

25 **(C) 3:2**

Let's think of the total time spent as t and the total distance traveled as d.

The speed for the first part of the journey can be expressed as: $\frac{\frac{3}{4}d}{\frac{2}{3}t} = \frac{3d}{4} \cdot \frac{3}{2t} = \frac{9d}{8t}$

The speed for the second part of the journey can be expressed as: $\frac{\frac{1}{4}d}{\frac{1}{3}t} = \frac{1d}{4} \cdot \frac{3}{1t} = \frac{3d}{4t}$

The ratio of the two speeds can be expressed as: $\frac{\frac{9d}{8t}}{\frac{3d}{4t}} = \frac{9}{8} \cdot \frac{4}{3} = \frac{3}{2} = 3:2$.

---

26 **(A)**

Picture A can be seen on the opposite side of the cube. From the first and the fourth cube it can be noted that the opposite face of the cube would have little marks (short segments) close to the center of the cube's face.

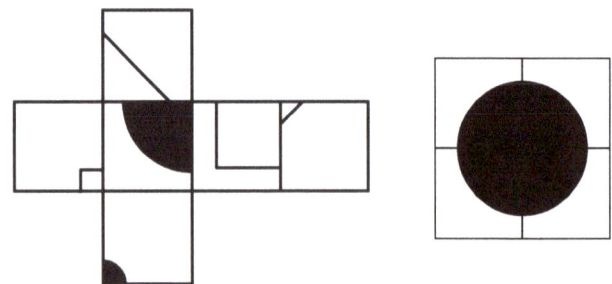

27 (B) 5

We don't know whether each of the damsels is telling the truth or lying at the beginning. However, from the last statement we know that there are at least 8 damsels, because only a lying damsel would answer "Yes." In the second statement, those damsels are telling the truth, and the other 4 "Yes" answers are serfs. For the first statement, we have 4 serfs and 8 lying damsels who will answer "Yes," so the other 5 answers are knights.

28 (C) 273

In order for numbers to be divisible by 13, the numbers must be multiples of 13. In order to keep the smallest possible value of M and only two numbers divisible by 2, we must include the two smallest even multiples of 13. The following set of numbers fulfills this condition: 13, 26, 39, 52, 65, 91, 117, 143, 169, 195, 221, 247, and 273. The smallest possible value of M is 273.

29 (A) 16

Note the image:

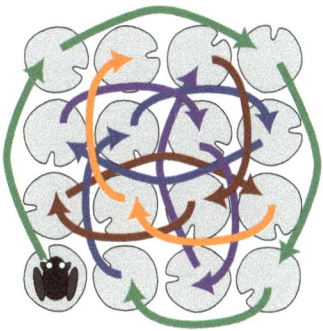

The frog can reach all 16 leaves.

30 (B) 5

The smallest number of such gray unit segments is 5.

2016

3 Point Solutions

1 **(C) 17**
3 is not included because it is lower than 3.17. Starting at 4 and counting upwards, stopping at 20 reveals that there are 17 numbers between 3.17 and 20.16.

2 **(A)**

By inspection, (A) has 4 axes of symmetry. (B) has 2, (C) has 3, (D) has none, and (E) has 1.

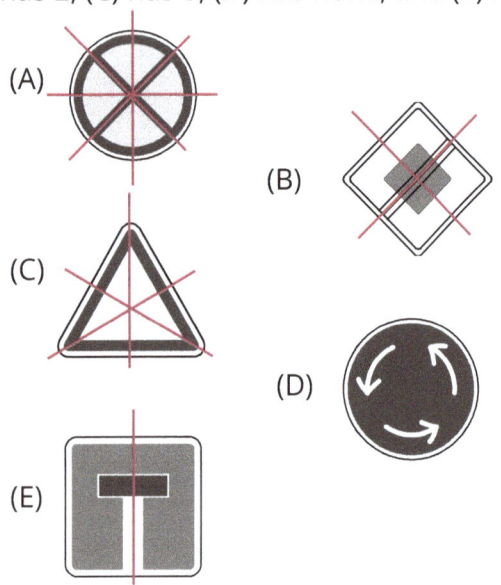

3 **(C) 270°**
Let α and β be the two acute angles of the triangle. Then the marked angles are 180° − α and 180° − β. Their sum is 360° − α − β, or 360° − (α + β). Since α and β are the other two angles in the right triangle, α + β = 90°. Therefore, the sum of the marked angles would be 360° − 90° = 270°.

4 **(D) 38**
Since Jenny subtracted 26 instead of adding, we must add 26 twice — once to undo the unintended subtraction and once again to perform the intended addition.
−14 + 26 + 26 = 38.

5 **(B)**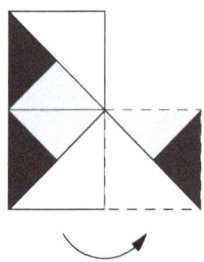

The dark shaded region is on the left after the first flip, then moves to the right after the second flip. The light shaded region is on the top after the first flip, then remains on top after the second flip.

6 **(A) 999**
555 groups × 9 stones ÷ 5 stones = 999 groups.

7 **(C) 9**
$45 \cdot \frac{100}{60}$ is 100% of teachers, so 12% of teachers is $45 \cdot \frac{100}{60} \cdot \frac{12}{100} = 9$.

8 **(C) 100**
Rotate each circle 90° to see the amount of shaded area is equal to the area of 4 triangles. The area of the four triangles is half the area of the rectangle, so 20 · 10 ÷ 2 = 100.

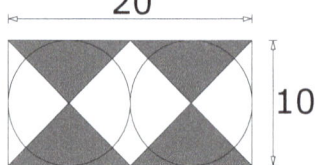

9 **(B) 8**
Let n be the number of parts of the 1 meter piece of rope. Then $2n$ is the number of parts of the 2 meter piece of rope. $3n$ is the number of all parts and $3n$ is a multiple of 3. Among the options shown, only 8 is not a multiple of 3.

10 **(C) 6**
The routes are SPQSRQ, SPQRSQ, SQPSRQ, SQRSPQ, SRQPSQ, and SRQSPQ.

4 Point Solutions

11 **(E) 32 cm**
Each rectangle shares a half of its perimeter with the perimeter of the square. The whole perimeter of the square consists of four such segments, so the perimeter of the square has the length of 4 × (16 ÷ 2) = 32 cm.

12 **(E) 40**
There is only 1 red bead among all beads, so there will be 1 red bead in a group of beads with 90% being blue. In that group, the 1 red bead is 10% of the beads, so the group has 10 beads. 9 of them are blue and you have to remove 40 blue beads to get 9 blue beads.

13 **(C) $\frac{29}{57}$**
Subtracting $\frac{1}{2}$ from each number and taking the absolute values, we get respectively
$\frac{29}{158}, \frac{5}{118}, \frac{1}{114}, \frac{25}{158},$ and $\frac{22}{184}$.
The third value is the smallest, so (C) is the closest to $\frac{1}{2}$.

14 **(B) Glen and Carl**
Since there are 8 players and this is a knock-out style tournament, 3 wins are required to win the tournament and 2 wins will get a player to the finals. The players with 2 and 3 wins are Glen and Carl.

15 **(B)**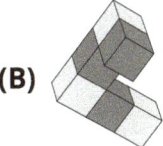
Refer to the solid on the right. Use your left hand. Put the thumb in the direction of the short column of cubes (the *y*-axis) and your open fingers vertically up along the adjacent column (the *x*-axis). When you curl your fingers they point into the direction of the 3rd column of cubes. Thus, the solid has the left-hand orientation. Rotations do not change the orientation. Among the five options shown only (B) has the right-hand orientation. Indeed, use your right hand. Put the thumb in the direction of the short column of cubes and your open fingers along the adjacent column. When you curl your fingers they point into the direction of the 3rd column of cubes, so the solid (B) has the right-hand orientation and can't be seen from any angle.

16 **(D) 89**
For each twin, add 3 to the sum of five ages to have a sum of five identical numbers. This new sum is a multiple of 5. Testing the options shown: (A) 36 + 6 = 42, (B) 53 + 6 = 59, (C) 76 + 6 = 82, (D) 89 + 6 = 95, and (E) 92 + 6 = 98. The only multiple of 5 is 95, so (D) is the only valid option. The triplets are 19 and the twins are 16.

17 **(D) 57 cm**
We can find the length of the strip by tracking its left edge. The total length of nine consecutive segments is (6 + 9 + 6 + 3 + 6 + 9 + 6 + 3 + 9) cm = 57 cm.

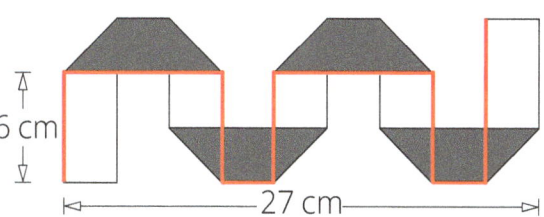

18 (B) 11
The distance Jum travels after N jumps is $6N$. The distance Per travels after each jump is $1 + 2 + \ldots + N - 1 + N = N \times (N + 1)/2$. The distances traveled will be equal when $6N = N(N + 1)/2$. Solving for N yields 11.

19 (D) 105
Each of the six outer dice is glued to the inner die (the center of the solid) in such a way that the two faces glued together have the same number of dots. Therefore, the sum of dots on the glued faces of the six outer dice is the same as the number of dots on the inner die. The number of dots on any standard die is $1 + 2 + 3 + 4 + 5 + 6 = 21$. Thus, the number of dots on the surface of the solid is $6 \cdot 21 - 21 = 5 \cdot 21 = 105$.

20 (B) 12
Let p be the number of mixed pairs. Then $3p$ is the number of boys and $2p$ is the number of girls. The number of all students is $3p + 2p$, so $5p = 20$ and $p = 4$. So, there are $3 \times 4 = 12$ boys.

5 Point Solutions

21 (D) 9
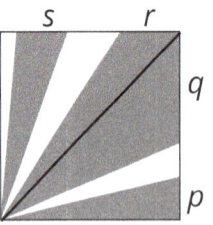
The shaded area is made up of four triangles with bases p, q, r, and s. For each triangle the height corresponding to the base is 6 since each side of the square has the length of 6. The total shaded area is $1/2 (6p + 6q + 6r + 6s) = 27$, so $p + q + r + s = 9$.

22 (D) 12:30
If Theo thinks it is 12:00, it is really 12:15 (5 + 10 = 15 minutes delay). If it is really 12:15, Leo thinks it is 12:30 (5 + 10 = 15 minutes ahead).

23 (E) 8
There were $1.5 \times 12 = 18$ cupcakes consumed and they were eaten by $12 - 2 = 10$ girls. For each of the 10 girls we subtract 1 cupcake, so there are $18 - 10 = 8$ cupcakes left. Nobody ate more than 2 cupcakes, so the 8 additional cupcakes were consumed by 8 girls. Hence, 8 of the girls ate two cupcakes.

24 (D) 7
Let T be the number of waffles she starts with and G the number of waffles she gives to each granny. View the process in chronological order. She starts with T waffles: $T \to T/2$ (wolf's 1st steal) $\to T/2 - G$ (1st granny) $\to (T/2 - G)/2 = T/4 - G/2$ (wolf's 2nd steal) $\to T/4 - 3G/2$ (2nd granny) $\to (T/4 - 3G/2)/2 = T/8 - 3G/4$ (wolf's 3rd steal) $\to T/8 - 7G/4 = 0$ (3rd granny). Thus, $T/8 = 7G/4$ and $T = 14G$.
T is always divisible by 7 and not necessarily by any of the other numbers.

25 (E) 17

2	1	2	
1	0	1	2
2	1	2	
	2		

Top layer — 11 +

	2		
2	1	2	
	2		

2nd layer — 5 +

	2		

3rd layer — 1 = 17

Each number shows on which day the corresponding cube changed to gray. Cross sections of the big cube show the top faces of the layers.

26 (C) 44
The numbers must be distinct; thus we seek two numbers multiplying to 225, but both should be relatively large. Obviously 15×15 doesn't work because they are the same. Thus we try 9×25. For the two smallest numbers, it could be 4×4. However these are again the same numbers. We try 2×8. Thus our numbers are 2, 8, 9, and 25, as there are no integers between 8 and 9, which sums to 44.

27 **(A)** *A*

For each vertex V, let R_V be the radius of the circle centered at V. For each side of the pentagon, two circles centered at the endpoints of the side are externally tangent, so $R_A + R_B = 16$, $R_B + R_C = 14$, $R_C + R_D = 17$, $R_D + R_E = 13$, $R_A + R_E = 14$.
Adding these equations together gives that $R_A + R_B + R_C + R_D + R_E = 37$, so $R_A = 37 - (R_B + R_C) - (R_D + R_E)$
$= 37 - 14 - 13 = 10$. Thus we can find that $R_B = 6$, $R_C = 8$, $R_D = 9$, $R_E = 4$. So, the longest is R_A.

---

28 **(E)** 118

Note that when computing the second layer, the center cube in the bottom layer is used 4 times, the corner cubes are each used once, and the middle cubes twice. We want to maximize $S = 4c + 2S_2 + S_3$ where c is the number at the center cube, S_2 is the sum of numbers at the centers of edges, S_3 is the sum of numbers at corners, and $c + S_2 + S_3 = 50$. Intuitively, we will maximize $4c + 2S_2 + S_3$ by putting the smallest four numbers in S_3, the medium four numbers in S_2 (they are multiplied by 2), and making c the biggest number to quadruple its contribution to S. Hence, $S_3 = 1 + 2 + 3 + 4 = 10$, $S_2 = 5 + 6 + 7 + 8 = 26$, and $c = 50 - 10 - 26 = 14$. 1, 2, 3, 4, 5, 6, 7, and 14 are different positive integers, their sum is 50, and $S = 4 \cdot 14 + 2 \cdot 26 + 10 = 118$.

29 **(C)** 17

Let us have cars $A, \ldots, E$ with $a + 1$, $b + 1, \ldots, e + 1$ passengers respectively. Here is the list of the number of neighbors for each passenger in a given car. $(A) = a + 1 + b$; $(B) = 1 + a + b + 1 + c$; $(C) = 1 + b + c + 1 + d$; $(D) = 1 + c + d + 1 + e$; $(E) = 1 + d + e$. Each of the numbers is either 5 or 10. $(B) = (A) + 1 + c$, so $(B) > (A)$, and so $(A) = 5$ and $(B) = 10$. $(D) > (E)$, so $(E) = 5$ and $(D) = 10$. $2 \cdot (A) = (B)$, so $2a + 2 + 2b = a + b + c + 2$. Hence, $c = a + b$. $(C) = 1 + b + c + 1 + d = 1 + b + (a + b) + 1 + d = (1 + b + a) + (b + 1 + d) = (A) + (b + 1 + d) > (A)$, so $(C) = 10$. $5 = (A) = a + 1 + b$, so $a + b = 4$ and $c = 4$. Similarly, $c = e + d = 4$. Therefore, the number of all passengers is $1 + a + 1 + b + 1 + c + 1 + d + 1 + e = (a + b) + c + (d + e) + 5 = 12 + 5 = 17$ if the requirements are consistent. Notice that $(B) = (C)$, so $a = d$. Also, $(C) = (D)$, so $b = e$. Consequently, the numbers of passengers in consecutive cars are: $1 + a$, $1 + b$, 5, $1 + a$, $1 + b$ where a, b are nonnegative integers such that $a + b = 4$. 1, 5, 5, 1, 5 is one of the options for distributing 17 passengers but there are 4 more such options.

30 (A)

Note that we must match pieces on the edges of the squares in order for them to be adjacent on the cube. In the 3rd picture of the true cube, note that we must have a white between two blacks. Squares 1, 2, and 4 can satisfy this property, but 5 doesn't. Thus 5 is opposite 3. Furthermore, we seek a square that has a white between two blacks, ruling out (D). Draw 3 in the center with 1, 2, and 4 surrounding it in the correct orientation. In the diagram, face 5 is at the center, the outer sides of the 1st, 2nd, and the 4th face must show the pattern b-w-b since the same faces are also adjacent to face 3 (in the big cube it is directly above face 5), and all sides of face 3 have the pattern b-w-b. This allows for only one configuration of the five given faces.

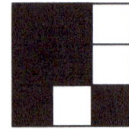

Note that we omit face 3 from the drawing, but the green missing face must have a b-w-b row on top. The green face's sides must match its left and right, so only (A) and (E) work. However, by counting black/white cubes (being careful to not double count), (A) gives 15 black cubes, while (E) gives at least 16 black cubes (assuming the central cube is white). Thus only (A) is valid.

2018

3 Point Solutions

1 (B) 19
Order of operations tells us to do parentheses first, so we get (20 + 18) ÷ (20 − 18) = 38 ÷ 2 = 19.

2 (E) TOOT
Each letter needs to have a vertical line of symmetry for the whole word to have a vertical line of symmetry when written with one letter on top of another. Of the options, only the word "TOOT" has a line of symmetry when written vertically.

3 (B) 9
A perimeter is the distance around a figure. A triangle with side lengths of 6, 10, and 11 has a perimeter of 6 + 10 + 11 = 27. An equilateral triangle has three sides of equal length. Therefore, if the perimeter of the triangle is 27, each side length would be 27 ÷ 3 = 9.

4 (D) 12
Let us solve the equation.
★ = (2 × 18 × 14)/(6 × 7) = 2 × 3 × 2 = 12.

5 (C)

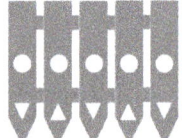

If the fence fell flat on the ground, he would see the back of the panel, appearing like a mirror image reflected around the bottom of the original panel. Make sure the answer includes the correct triangles and that the circles are positioned at the correct height on the fence.

6 (D) 20
The combined height of the steps needs to equal the vertical distance between the first and second floors, which is 3 m or 300 cm. If each step has a height of 15 cm, we can calculate the number of steps needed. 300 cm ÷ 15 cm = 20. Therefore, he needs 20 steps.

7 (C) 4
There are 4 ways:

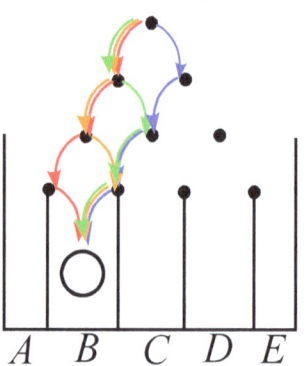

8 (C) 76 cm
Since the longer side of each small rectangle is 10 cm, this makes the longer side of the large rectangle 20 cm. Based on the image, the short side of five small rectangles is equivalent to the longer side of the large rectangle. This makes the short side of the small rectangles 20 ÷ 5 = 4 cm each. Knowing this, we can calculate the perimeter to be 10 + 10 + 4 + 10 + 4 + 10 + 10 + 4 + 10 + 4 = 76 cm.

9 (D) 4
We can see that the diameter of one circle is 7, making the radius 7 ÷ 2 = 3.5. This means that the distance between the side of the rectangle and the center of the circle is 3.5. Since there are two circles, it means that out of the 11 units that make up the rectangle, 7 units are between the sides of the rectangle and the centers of the circles. This means that the remaining length, the distance between the centers of the two circles, is 11 − 7 = 4.

10. (D) 2 cm

If both sides of the square are 3 cm, the area is 3 × 3 = 9 cm². The square is split into three equal pieces, so the area of each piece is 3 cm². Since we know that DC is 3 cm, the length DM can be found by using the formula for the area of a triangle, Area = $\frac{1}{2}$ × base × height. 3 = $\frac{1}{2}$ (3) × DM. DM is equal to 2 cm.

4 Point Solutions

11. (B) 6

The product is 300-something, and the second number is 20-something, so the first number needs to have 1 in the tens place. Anything more than that will give us a product greater than needed because it would be at least 23 × 20. So, in order for the answer to be a three-digit number that starts with 3, the first digit of the first number has to be 1, making the number 13. 13 had to be multiplied by a number that ends with 4 in order to give a product that ends with 2. This means the second number is 24. 13 × 24 = 312, so the third scribbled-out digit is 1. 1 + 4 + 1 = 6.

12. (C) 32

In order for there to be a middle row, there must be an odd number of rows. The only odd divisor of 40, other than 1, is 5. Therefore, there are 5 rows in total. With 5 rows, there are exactly 40 ÷ 5 = 8 squares in each row. If Andrew only colored the middle row, there will be 40 − 8 = 32 squares left uncolored.

13. (D) 20

If Philip weighs multiple copies of the book, he can divide the total weight by the number of copies to get the required precision. Since 10 grams is 20 times 0.5 gram, he would need at least 20 copies to find the weight.

14. (C) room 3

We know that the note on door 3 is false, as 2 + 3 does not equal 2 × 3. If the note on door 1 ("The lion is here.") was true, that would mean that the note on door 2 ("The lion is not here.") was true too. There is only one true statement, so the only note that can be true is the one on door 2. If the note on door 1 is false and the note on door 2 is true, this means that the lion has to be behind door 3.

15. (A) 11°

Let's mark all points which create triangles in the rectangle, as shown in the picture. The sum of the measures of all three angles in a triangle is 180°. We know that since one angle of triangle ABC is 26° and the angle at vertex B is 90°, the remaining angle of this triangle, at vertex C, is 180° − (90° + 26°) = 64°. The 64° angle, 33° angle, and the other angle at vertex C combine to form a semicircle, so their sum is 180°.

In triangle CDE, the angle at vertex C is 180° − (64° + 33°) = 83°. The two angles in triangle EFG are 90° and 10°, so the third angle at vertex E is 180° − (90° +10°) = 80°. The 80° angle, the 14° angle, and the remaining angle at point E form a semicircle. Knowing that their sum equals 180°, we can calculate the third angle to be 180° − (80° + 14°) = 86°.

This gives us two angles to triangle CDE, which contains angle θ. We can then calculate the remaining angle to be 180° − (86° + 83°) = 11°.

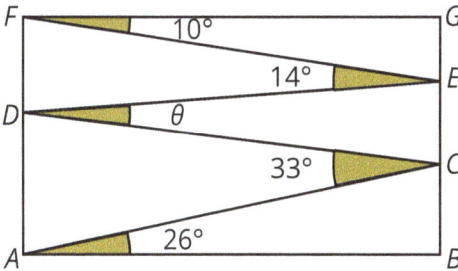

16 (D) 41
Remember that 1 is neither prime nor composite. Since 4 is not prime and any number that ends in 4 is also not prime, it needs to be used as a tens digit. We can use 1 for the ones digit to make 41. If we use 3 as the ones digit to make 43, we cannot make a prime number using the 1 and any of the remaining digits. Therefore, 41 must be on the list.

17 (D) 32
There are 365 days in 2018. According to the ad, there will be 365 − 350 = 15 days of no sun. The worst case scenario is if Willi's stay begins with sun and alternates days of sun and no sun. In this case, Willi would have to stay 32 days before he would see two consecutive days of sun.

18 (B) 20 cm²
Triangles A and B make up exactly half of the area of the rectangle. Since we know the sum of the two triangles to be 10 cm², the area of the rectangle is 20 cm².

19 (A) 17
No matter what order James writes the numbers in, the sum of the rows will always equal 1 + 2 + 3 + 4 + 5 + 6 + 7 + 8 + 9 = 45, and the sum of the columns will, too. This means that the sum of the rows and columns will equal 90. Since the five answers given add up to 73, the last answer must be 90 − 73 = 17.

20 (B) 2
The sum of the distances between the first point and the other points is 2018, while the sum of the distances between the second point and the other points is 2000. The difference in the sums of the distances between the first point and the other points and the distance between the second point and the other points is 2018 − 2000 = 18. Since there are nine total sums between the first point and the other points (excluding the second point), and the second point is closer than the first point to each of the other nine points by the same distance, the distance between the first point and the second point can be written using the equation $9x = 18$, so $x = 2$.

5 Point Solutions

21 (E) 17
There have been 24 + 29 + 37 = 90 votes placed so far. This means that 130 − 90 = 40 students still have to vote. We can now try the answers starting from the largest. If Akmal receives 17 more votes, this would bring him to 54 votes, with 40 − 17 = 23 votes remaining. Even if all 23 of those votes went to the candidate in 2nd place, Khairul, it would only put him at 52 votes, two fewer than Akmal. If Akmal got 16 more votes and the rest went to Khairul, the results would be exactly even, so the answer must be 17.

22 (C) 80 cm³
Let l represent the length, w represent the width, and h represent the height of the box in cm. From the picture, we can determine $2l + 2w = 26$, $w + h = 7$, and $l + h = 10$. Divide the first equation by 2: $l + w = 13$. Add all 3 equations up: $2l + 2w + 2h = 30$; then divide by 2: $l + w + h = 15$.

We can subtract each of the three original equations from this last equation to isolate variables:
$l + w + h − (l + w) = 15 − 13$ → $h = 2$
$l + w + h − (l + h) = 15 − 10$ → $w = 5$
$l + w + h − (w + h) = 15 − 7$ → $l = 8$

Since the volume of a rectangular box is given by the formula length × width × height, we can determine the volume of this box to be 5 × 2 × 8 = 80 cm³.

23 (B) 7

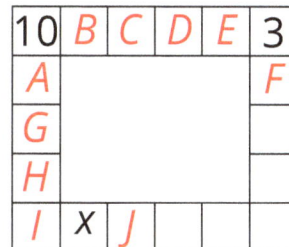

First work from 10 to 3 to determine B:
1) $10 + C = B$ → $C = B - 10$
2) $B + D = C$ → $D = C - B = B - 10 - B = -10$
3) $C + E = D$ → $E = D + 3 = -10 + 3 = -7$
4) $D + 3 = E$ → $C = D - E = -10 - (-7) = -3$
5) $B = 10 + C = 10 - 3 = 7$

Then work counterclockwise from 10 to determine x:
6) $A + B = 10$ → $A = 10 - B = 10 - 7 = 3$
7) $10 + G = A$ → $G = A - 10 = 3 - 10 = -7$
8) $A + H = G$ → $H = G - A = -7 - 3 = -10$
9) $G + I = H$ → $I = H - G = -10 - (-7) = -3$
10) $H + x = I$ → $x = I - H = -3 - (-10) = 7$
11) $x = 7$

---

24 (B) 40 m

Ian swam the 50 m length of the pool six times, meaning he swam a total of 300 m. Simon ran around the pool five times. We know that the length of the pool is 50 m, but we don't know the width, so the distance he ran, which is the perimeter of the pool, in m, can be represented by $5(2x + 100)$, where x is the width of the pool. We can then set the distance Ian swam equal to the distance Simon ran, dividing Simon's distance by his speed, r, and Ian's distance by his speed, 3r.

$$\frac{300}{r} = \frac{5(2x + 100)}{3r}$$

Simplifying the equation, we have:

$900 = 10x + 500$

Solving for x, we get:

$10x = 400$
$x = 40$

25 (D) 24 cm × 16 cm

If we rearrange the red dove, we see that it takes up 12 blocks and is half the flag. Each block, then, has an area of 192 ÷ 12 = 16 cm². The blocks are the squares of a grid, so each block has a side length of 4 cm. The flag is 6 blocks by 4 blocks, so its dimensions are 6 × 4 cm by 4 × 4 cm, or 24 cm by 16 cm.

---

26 (C) 3

The first move would be to take the domino with three and one dots and switch it with the domino that has six and one dots. The second move would be to take the domino with four and two dots and switch it with the domino that has three and one dots. The third and final move would be to rotate the domino that has three and one dots.

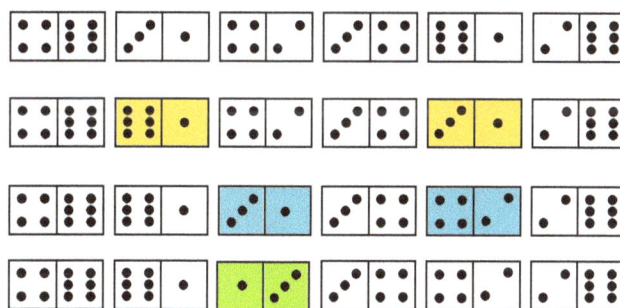

There is no solution for fewer than 3 moves. Exactly two dominoes have a 3, and both 3s are on the left side. Similarly, exactly two dominoes have a 1, and both 1s are on the right side. This means a domino with a 1 and a domino with a 3 must be flipped. There is a domino that has both 1 and 3, so we can flip this one to make the necessary adjustment in just 1 move. There are five mismatched pairs between the dominoes. The maximum number of matches we can create with a single swap is four because each domino has two sides, and by swapping we can change up to four sides in total. Therefore, at least 2 swaps are needed to fix all mismatches. This brings the total number of moves required to at least 3.

27. (B) 12

Since triangle *ABC* is equilateral, angles *A*, *B*, *C* are 60°. This means triangles *CNM*, *ALN*, and *BLM* are all 30°–60°–90°, and angles *CNM*, *LMB*, and *ALN* are 30°, while angles *MCN*, *LAN*, and *MBL* are 60°. This tells us that the side lengths are in the proportions $1:2:\sqrt{3}$. Let the length of each segment *CM*, *NA*, and *LB* be represented by a, so that the length of segments *CN*, *AL*, and *BM* can be represented by $2a$, and the length of segments *NM*, *NL*, and *LM* be represented by $\sqrt{3}a$. This means that each side of triangle *ABC* is $3a$.

Now let's use the formula for the area of an equilateral triangle,
$$A = \frac{\sqrt{3}}{4}s^2.$$

For triangle *ABC*, the area can be represented as
$$\frac{\sqrt{3}}{4}(3a)^2 = \frac{\sqrt{3}}{4}(9a^2) = 9\left(\frac{\sqrt{3}}{4}a^2\right).$$

For triangle *LMN*, the area can be represented as
$$\frac{\sqrt{3}}{4}(\sqrt{3}a)^2 = \frac{\sqrt{3}}{4}(3a^2) = 3\left(\frac{\sqrt{3}}{4}a^2\right).$$

This means that the area of triangle *ABC*, which we know is 36, is three times greater than the area of triangle *LMN*, so this area must be $36 \div 3 = 12$.

28. (E) 32

Let x represent the amount Choo spent, $0.15x$ represent the amount Bustan spent, and $1.6x$ represent the amount Azmi spent. The equation representing the total amount spent is:

$$x + 0.15x + 1.6x = 55.$$

Combining like terms, we have:

$$2.75x = 55.$$

To solve for x, we divide both sides of the equation by 2.75:
$$x = \frac{55}{2.75}$$

Simplifying the right side, we find:

$$x = 20.$$

Therefore, Choo spent $20. To find the amount Azmi spent, we can multiply this amount by 1.6:

$$1.6 \times 20 = 32.$$

So, Azmi spent $32.

(C) 4.01 m

Viola's current average distance can be represented by the equation

$$\frac{3.8x + 3.99}{x + 1} = 3.81$$

where x is the number of times she has jumped so far. Solving algebraically, we find that $x = 18$. To determine the distance Viola needs to achieve on her next jump, we can use the equation

$$\frac{3.8x + 3.99 + y}{x + 2} = 3.82$$

where y is the distance she needs on her next jump. Substituting $x = 18$ into the equation, we can solve for y.

$$\frac{3.8(18) + 3.99 + y}{18 + 2} = 3.82$$

Simplifying the equation, we have:

$$\frac{68.4 + 3.99 + y}{20} = 3.82$$

Multiplying both sides of the equation by 20, we get:

$68.4 + 3.99 + y = 3.82 \times 20$.

Simplifying further, we have:

$68.4 + 3.99 + y = 76.4$.

Combining like terms, we find:

$72.39 + y = 76.4$.

Subtracting 72.39 from both sides of the equation, we get:

$y = 76.4 - 72.39$.

Calculating the result, we find:

$y = 4.01$.

Therefore, Viola needs to achieve a distance of 4.01 on her next jump.

 (C) 36°

Triangle ABC is an isosceles triangle. We are given that $AK = LB$, so KB must equal LC. Because there are two pairs of adjacent sides of the same length, we can determine that $AKLC$ is a kite. $\angle A$ and $\angle C$ are both equal, and using the properties of a kite, we can determine $\angle KLC$ to be equal to $\angle A$ and $\angle C$, as well. Let $\angle A$, $\angle C$, and $\angle KLC$ each be represented x. The measures of the four angles of a quadrilateral add up to 360°. Since we determined $\angle A$, $\angle C$, and $\angle KLC$ to be equal, we can determine $\angle AKL$ to be $360° - 3x$.

In triangle BKL, BL and KL are equal, meaning $\angle B$ and $\angle K$ are equal. Since $\angle BLK$ and $\angle KLC$ make up a linear pair, we can subtract the measure of $\angle KLC$ from 180° to find $\angle BLK$. $\angle BKL$ and $\angle AKL$ also make up a linear pair, so the measure of $\angle BKL$ can be written as $180° - (360° - 3x)$, which means $\angle B = 180° + 3x$.

We know that in triangle BKL, all the angles ($\angle B$, $\angle BKL$, and $\angle BLK$) add up to 180°. This can be represented by using $\angle B + \angle BKL + (180° - \angle KLC) = 180°$. If we solve this, we can get $\angle B = x/2$. Earlier, we determined $\angle B$ to also equal $180° + 3x$.

We can now set $x/2$ and $180° + 3x$ equal to each other to solve for x. This gives x (the measures of $\angle A$, $\angle C$, and $\angle KLC$) to equal 72°. Since $\angle ABC$ is represented by $\angle B$ and that's the angle we are looking for, we can plug 72° for x into $\angle B = x/2$, giving a final answer of $\angle B = 36°$.

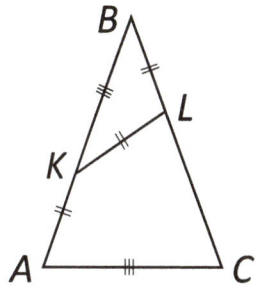

2020

3 Point Solutions

1 (B) 1
2 is a prime number. Since all the other numbers are even (they end in 2 or 0), they cannot be prime because they have 2 as one of their divisors.

2 (A)

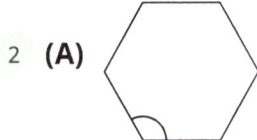

One can find the angle a of a regular polygon with n sides by the formula $a = \dfrac{180(n-2)}{n}$ which increases as n increases. Then the hexagon, which has the greatest n number of sides of the figures shown, has the greatest interior angle.

3 (C) 6
Miguel solves 6 × 4 = 24 problems in 4 days. Lazaro takes 24 ÷ 4 = 6 days to solve them.

4 (A) $\dfrac{8+5}{3}$

Answer (E) is less than 1, and both (B) and (C) are equal to 1. (A) is 11/3 and (D) is 11/5. The smaller denominator gives a larger answer.

5 (E) $\dfrac{1}{2}$

The large square is divided into fourths. The top left fourth has half of it shaded, or an eighth. The top right fourth also has half of it shaded, for another eighth. The bottom right fourth is completely shaded.
$\dfrac{1}{8} + \dfrac{1}{8} + \dfrac{1}{4} = \dfrac{1}{2}$, which is (E).

6 (E) 8
Since there are 4 teams, each team will play 3 games. The highest score possible comes from 3 wins, which give 3 × 3 = 9. The second-highest score possible is with two wins and a tie, 3 + 3 + 1 = 7, or (D). Then 8, (E), is not a possible score. Though not required, the other scores can be shown to be possible: (A) with a win, a tie, and a loss, (B) with a win and two ties, and (C) with two wins and a loss.

7 (D) 18
A hexagon can be made by adding the triangles as shown. This takes 18 triangles.

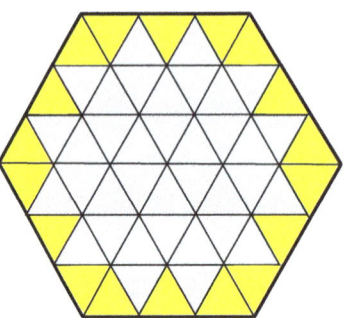

8 (B) −120
The solution should be negative, so we need either 3 negative numbers or 1 negative and 2 positives. We should take the largest such numbers possible. Looking at the absolute values, with 3 negatives, we can get 5 × 3 × 1 = 15, or (E). With 1 negative and 2 positives, we can take 5 × 6 × 4 = 120, or (B). Since these are the three numbers with the largest absolute values, the answer is (B).

9 (D) 5 hours
Going by bus both ways takes John 1 hour, so going one way takes 0.5 hours. Since going by bus one way and walking the other takes 3 hours, walking one way takes 3 − 0.5 = 2.5 hours. Then, walking both ways takes 2.5 + 2.5 = 5 hours.

10 (B) 43

The sum of the rows and the sum of the columns are both the sum of the numbers in all 9 cells, so they must be equal. 24 + 26 + 40 = 90, and 90 − 27 − 20 = 43 (B). Though not necessary, one possible arrangement of numbers is shown.

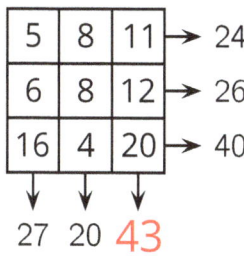

4 Point Solutions

11 (A) 1 km

Since the left signpost is 2 km from Atown and the right signpost is 7 km from Atown, the signposts are 7 − 2 = 5 km apart. The directions of the Betown arrows tell us that Betown is between the signposts, so the sum of the distances from the signposts to Betown should be the total distance between the signposts, or 5 km. Since the left signpost is 4 km away from Betown, the right signpost is 5 − 4 = 1 km away.

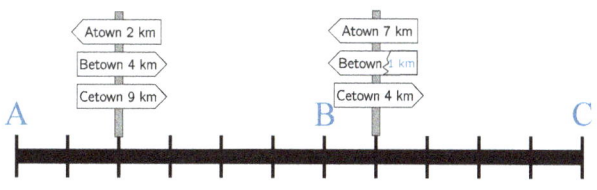

12 (C) 4 km

March has 31 days. Since Anna wants to average 5 km of walking each day, she needs to walk 31 × 5 = 155 km total. She has already walked 95 km, so she has 155 − 95 = 60 km left to walk. Starting on March 17, there are 15 days through March 31, inclusive. Then she needs to walk 60 ÷ 15 = 4 km a day.

13 (B)

From the image, we see that in clockwise order, we want the four diagonal pieces to be black, black, white, and gray. Then, starting with the gray side between the two black diagonal pieces, in clockwise order we want the sides to be gray, black, gray, and white. The only answer that shows this arrangement is (B).

14 (C) 25

If we count all the students who swim, and all who dance, we are double-counting the students who do both.

$\frac{3}{5} + \frac{3}{5} = \frac{6}{5}$, so $\frac{6}{5} - 1 = \frac{1}{5}$ of the students are double-counted. Then 5 students are $\frac{1}{5}$ of the total, so there are 25 students in all.

15 (C) 24

By splitting the segments as shown, we can see that $a = a_1 + a_2 + a_3$, $b = b_1 + b_2$, and c is repeated twice. Thus, the perimeter is $2(a + b + c) = 2(3 + 5 + 4) = 24$.

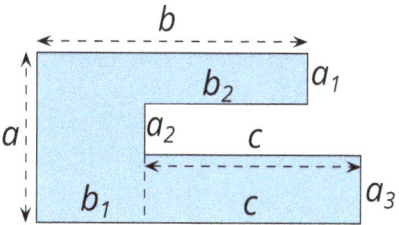

16 (C) 4

The small cubes at the corners of the large cube each have three faces exposed, so at least one of these faces must be white, as shown on the left cube. Looking at the two opposite corners filled in on the left cube, we see that at least two faces must be incomplete. However, by choosing four corners to leave the same face blank, we can have an entire strip of red around the cube as in the right picture, achieving four complete faces. The faces opposite the top and front are also red, and the face opposite the white face is also incomplete.

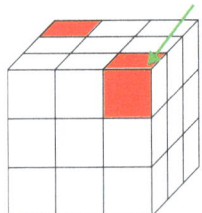
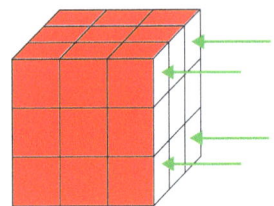

17 (A) 1 cm²

Each of the cross-hatched areas is half of a rectangle. The corners of the cross-hatched figure are made from opposite corners of congruent right triangles, thus they are all right angles, so the figure is a square. The area of the cross-hatched figure, with the white square inside, is 5 cm × 5 cm = 25 cm². This is equal to half the area of the rectangles plus the area of the small square. Adding another half of the area of the rectangles gets back the large square, with area 49 cm². Then the area of the rectangles is (49 cm² − 25 cm²) × 2 = 48 cm², and the area of the small square is 49 cm² − 48 cm² = 1 cm².

18 (D) 400%

Since 20% is equivalent to 1/5, the boss's salary is 5 times larger than Werner's. Then Werner needs to increase his salary fivefold, which requires an additional 4 times as much as he makes, or 400%.

19 (B) 24

Irene could be looking from the north, east, south, or west. In each row that she sees, to maximize the blocks, each tower of cubes should have the maximum possible amount. Then looking from the north, there are 3 towers of 2 cubes each, 2 towers of 4 cubes each, 1 tower of 1 cube, and 3 towers of 3 cubes each. This yields 3 × 2 + 2 × 4 + 1 × 1 + 3 × 3 = 24 cubes. Making a similar calculation yields 22 cubes from the east, 21 cubes from the south, and 23 cubes from the west. So the answer is (B). Note: Before counting, we could try and make sure we are getting as many 4-cube towers and as few 1-cube towers as possible, which happens when looking from the north. This arrangement also has the maximum number of 3-cube towers.

20 (E) 3, 4, 2, 1, 5

Each answer has 3 on top. We can attempt to draw examples of each fold, and see that in (E), the situation is impossible. One thing to notice about (A)–(D) is that all four answers group 1, 2 and 4, 5 together. As shown in (A) and (B), with 3 on top, 4 and 5 can be folded either as 4, 5 or 5, 4. Also, (C) and (D) show a similar situation with 1 and 2. Then either group can be folded in first towards 3, for example the as in the switched order between (B) and (C).

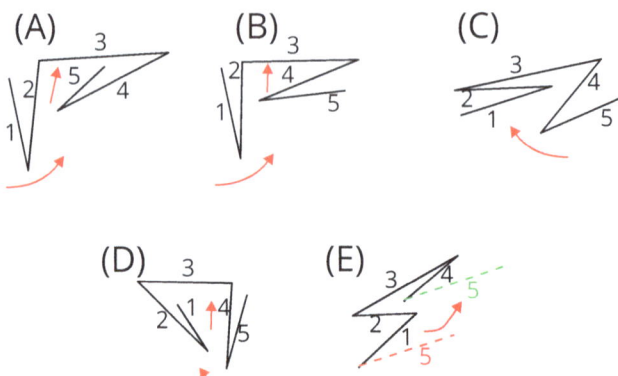

5 Point Solutions

21 **(A) green**
Since there are 3 consecutive red cubes, one of the ends is red, and the 10th cube from the left is blue, the red cubes sit on the left (otherwise the 12th, 11th, and 10th cubes would be red). Then there is an arrangement that looks like RRR??????B?Y. The 4 green cubes are all together, so they must come before the blue cube. But in order to fit, the green cubes can take up positions 4, 5, 6, 7; 5, 6, 7, 8; or 6, 7, 8, 9. In all these cases, the 6th cube is green.

22 **(A) 112.5°**
By folding the paper in, Zaida cuts the corner into four congruent angles. Each of the angles marked "o" are equal to 90° ÷ 4 = 22.5°. The top-left angle is composed of two of these "o" angles and is therefore 22.5° + 22.5° = 45°. The largest angle in the quadrilateral is one of the two congruent angles greater than 90. Then this is half of 360° − (90° + 45°) = 225°, or 112.5°.

23 **(D) 10**
If half of the number is divisible by 2, then the number must be divisible by 4. Similarly, the number must also be divisible by 9 and by 25. Since none of 4, 9, or 25 share any factors, the number must be divisible by all of them, so it must be a multiple of 4 × 9 × 25 = 900. The four-digit multiples of 900 are 900 × 2, 900 × 3, ..., 900 × 11, which gives 10 numbers.

24 **(B) 1**
First, Judge III must give Berta 3 points in order to sum to 5. Next, Emil must have scores of 4, 4, and 3 in order to add up to 11. Judge III must give Emil a 4 since he gave Berta 3. Let's assume that Judge II gives Emil a 4 as well. Then Judge I must give Emil a 3 and must give a 4 to either Clara or David. Clara cannot have a 4 since her sum is 3. If David gets a 4 from Judge I, he must have a 0 from Judge II, which he cannot since Judge II gave a 0 to Clara. Then Judge II cannot give Emil a 4, so he must give him a 3. Then Adam cannot have a 3 from either Judge II or III, and thus his two missing scores must be a 1 and a 4 in order to sum to 7. Since Judge III gave Emil a 4, he will give Adam a 1.

A possible solution is given:

	Adam	Berta	Clara	David	Emil
I	2	0			4
II	4	2	0		3
III	1	3			4
Sum	7	5	3	4	11

	Adam	Berta	Clara	David	Emil
I	2	0	1	3	4
II	4	2	0	1	3
III	1	3	2	0	4
Sum	7	5	3	4	11

25 **(C) 8**
If the numbers at the edges are A, B, C, D, then Saniya adds up $AB + BC + CD + DA = A(B + D) + C(B + D) = (A + C)(B + D) = 15$. Since A, B, C, D are positive, $A + C > 1$ and $B + D > 1$. Then the two sums are 3 and 5, and $A + B + C + D = 3 + 5 = 8$.

26 **(C) 8**
Assume the triangles have legs of length 1. Two isosceles right triangles make a square of area 1, and we can create larger squares by having a square number of these triangle pairs. So we want $2m^2 \leq 52$, or $m^2 \leq 26$. This gives us squares with sides of 1, 2, 3, 4, and 5 legs. However, the triangles can also be combined along the legs into larger tiles of 4 triangles each, with the hypotenuses as sides. Here, $4n^2 \leq 52$, or $n^2 \leq 13$. This gives us squares with sides of 1, 2, and 3 hypotenuses. Any other square would have a combination of legs and hypotenuses for sides. Since the length of the hypotenuse can be found by the Pythagorean theorem to be $\sqrt{2}$, the area of the square would have to be $(a + b\sqrt{2})^2 = a^2 + 2b^2 + 2ab\sqrt{2}$, which is irrational unless a or b is 0. Since each triangle has a rational area of 1/2, we have found all the possible squares. Then the answer is 5 + 3 = 8.

---

27 **(E) 96**
Each sphere will rest on four spheres in the layer below it. So each of the spheres above the base has four contact points (black), for $4(9 + 4 + 1) = 56$ contact points. Additionally, the spheres in each layer touch their vertical and horizontal neighbors (red). For a square of size n, this gives n rows of $(n - 1)$ contact points horizontally and n columns of $(n - 1)$ contact points vertically. Then we must add $2(4 \times 3 + 3 \times 2 + 2 \times 1) = 40$ points to the 56 above, or $40 + 56 = 96$ total.

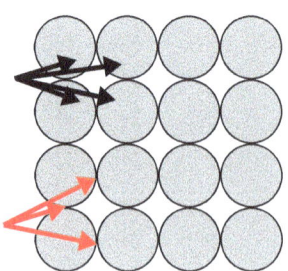

28 **(D) 20 m**
If all four children come to the trainer, the sum of their walks will be the perimeter of the pool. This can be seen in the first two pictures, where the two children on the same side as the trainer make a walk equal to that side, while the two children opposite do the same after each walking the length of one of the perpendicular sides. Since the children waled 50 m, the trainer must then walk 2(25 m + 10 m) − 50 m = 20 m to get to the last child.

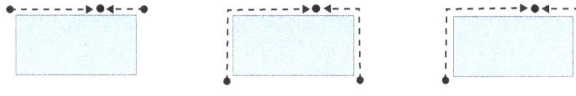

---

29 **(D) 165 m**
When Boris ran the last 15 m, Carl ran 35 − 22 = 13 m. Then every time Boris runs 15 meters, he moves 2 meters ahead of Carl. By the end of the race, Boris was 22 meters ahead. Then $\frac{22}{2} = \frac{x}{15}$, and solving for x we get 165 m.

---

30 **(C) 3**
From the fifth clue, we know that 2, 4, 6, and 7 are not in the four-digit number. Then, from the first clue, we know that 1 and 3 must be in the number. From the second clue, we know that either 9 or 8 is a digit of the number. If that digit were 8 and not 9, then the third clue implies that 5 and 0 are in the number. But then the four-digit number must include 1, 3, 8, 5, and 0, which is one digit too many. So from the second clue, we know the first digit is 9. Then in the third clue, 9 is the digit in the wrong place, and 0 is correctly in the second spot, since 9 must be first so 5 cannot be right. Finally, from the first clue, we know the 3 is not in the third spot, so the number must be 9013. The last digit is 3.

2022

2022

3 Point Solutions

1 **(E) 2, 3, and 5**
Follow the lines and add more arrows. Sanath paddles in a clockwise direction around buoys 2, 3, and 5.

2 **(B)** $\boxed{8}$
Because any arrangement of the pieces will have the same number of digits, the smallest number has the smallest digits as far left as possible. Because 1 < 3 < 4 < 5 < 8, the smallest possible number that can be made with the five pieces is:

| 107 | 31 | 4 | 59 | 8 |

3 **(C) 84**
During every routine Kengu advances 9 units. So after 9 routines he ends up at 81. After that he will land at 81 + 3 = 84. The next jump will bring him to 84 + 3 = 87, which is already bigger than the answers listed.

4 **(B)** $\boxed{60 \text{ HOH } 09}$

Flip the page, and notice that

$\boxed{60 \text{ HOH } 09}$

remains the same.

5 **(C) 4 × 8 × 12**
Let the dimensions of the brick be L, W, and H, where L goes from left to right, W goes into and out of the page, and H goes up and down. Because the bricks form a cube, we know that 3L = 2W = 6H, or, L = 2H and W = 3H. Therefore, H is the shortest dimension (4 cm). So, the side lengths of the cube are: H = 4 cm, L = 2 × 4 = 8 cm, and W = 3 × 4 = 12 cm.

6 **(A)**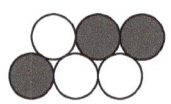

Note that the caterpillar's segments alternate black and white colors, and the 2 edges are black and white in color. For every option, start with any colored circle and follow to the next circle, which must be of the opposite color. Repeat this until you complete all six circles, or cannot find an adjacent circle of the opposite color. The only path that correctly matches all 6 segments is option (A).

7 **(D) between 15 and 18**
The sum of all the numbers is 6 + 9 + 12 + 15 + 18 + 21 = 81. This means that we counted 81 − 45 = 36 too much. If we flip the sign before one of the numbers n to a minus sign, the sum will decrease by $2n$. So, the minus sign should be before 36 ÷ 2 = 18. Check that 6 + 9 + 12 + 15 − 18 + 21 = 45.

8 **(B) B**
Look at path 1 from the bottom of the picture to the top: at the left hand side there are 2 trees and at the right hand side there are 3 trees. So, the new tree must be to the left of that path. Look at path 2 that starts high on the left hand side of the picture and ends low at the right hand side: above this path there are 2 trees and below this path there are 3 trees. So, the new tree must be above this path. Using both observations, the new tree must be planted in region B; this placement also works with the third path.

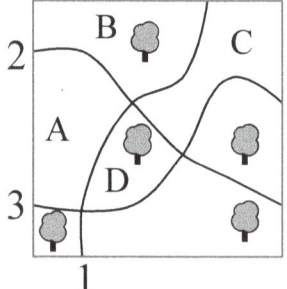

9. (A) 25
The odd digits are 1, 3, 5, 7, and 9. For the first digit, we can only use '1'; for the second and third digits, we can use all 5 possible odd digits. Hence, the total number of integers between 100 and 300 with only odd digits is $1 \times 5 \times 5 = 25$.

---

10. (C) 5
The last digit of the square of a number is determined by the last digit of the number. We know that the last digit of the second number is 2, meaning that its square will end with the digit 4. Because the sum of the two squares ends with a 9, we know that the square of the first number must end with a 5, meaning that the first number itself must also end with a 5. Notice that $2385^2 + 1202^2 = 7133029$, so there is at least one complete solution to the problem.

4 Point Solutions

11. (D) 6
A stack of 8 glasses is 42 cm tall and a stack of 2 glasses is 18 cm tall, so the extra 6 glasses add 24 cm to the height. This means that each extra glass added to the stack increases its height by $24 \div 6 = 4$ cm. If 4 glasses are added to the stack of 2 glasses, they would create a stack of height 18 cm + (4 × 4 cm) = 34 cm. If we add another glass (for a total of 7 glasses), the height would be 34 + 4 = 38 cm, which is greater than the shelf height (36 cm). This means that a stack of 6 glasses is the largest that can fit ion the shelf.

---

12. (D) 58
Since the sum of the dots on opposite faces of a standard dice is always 7, the total number of dots on the front face and the back face of the arrangement equals $4 \times 7 = 28$. The same is true for the top face and the bottom face of the arrangement. So, only the left face and the right face control the number of dots on the whole surface. To minimize that number, arrange the visible left and right faces to each have one dot. Therefore, the arrangement with the minimal number of dots on the entire surface is 28 + 28 + 1 + 1 = 58 (front and back, top and bottom, left, and right, respectively).

---

13. (E) 16
Let a, b, and c be the ages of the three sisters. Based on the averages, $(a + b + c) \div 3 = 10$, $(a + b) \div 2 = 11$, and $(a + c) \div 2 = 12$. Therefore, $a + b + c = 30$ (1), $a + b = 22$ (2), and $a + c = 24$ (3). Subtracting (2) from (1), we get $c = 8$. Subtracting (3) from (1), we get $b = 6$. Substituting for b in (2), we get $a = 16$. The eldest sister is therefore 16 years old.

---

14. (A) 48 m²
Draw a line from the lower left-hand corner to the upper right-hand corner of the flowerbed. The midpoint P of this segment is the center of the square. So its distance from any side of the square is 12 m ÷ 2 = 6 m. This segment divides the two quadrilaterals of daisies into four congruent triangles. The triangle PEB has area 1/2 × base × height = 1/2 × 4 m × 6 m = 12 m². Therefore, all four triangles have a total area of 4 × 12 m² = 48 m².

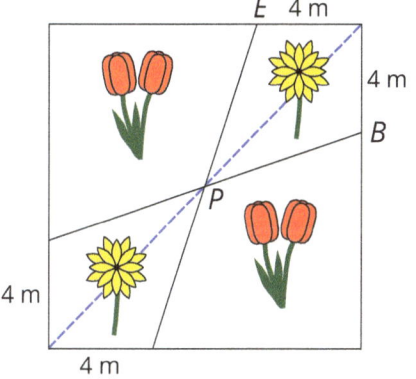

---

15. (C) 3:40 p.m.
If one clock gains one minute every hour and the other loses two minutes every hour then the time difference increases by 3 minutes per hour. So if the time difference is 60 minutes, the last time that the clocks were set to the correct time is 20 hours ago. During that period, the faster clock advanced an additional 20 minutes and it shows 12:00 noon, but the actual time is 11:40 a.m. 20 hours ago, it was 11:40 a.m. −12 hours = 11:40 p.m., then 11:40 p.m. − 8 hours = 3:40 p.m.

16 (B) 8
The sum of all Werner's numbers and all Ria's numbers together is 22 + 34 = 56. But when taken in pairs, Werner's number N plus Ria's corresponding number $7 - N$, the sum of each pair is $N + (7 - N) = 7$. So there are $56 \div 7 = 8$ pairs. Werner wrote 8 numbers on the paper.

---

17 (D) 17
The numbers 30 and 105 are multiples of 5, so 5 must be placed at the intersection of those lines. Similarly, the numbers 105 and 28 are multiples of 7, so 7 must be placed at the intersection of those lines. It is now easy to complete the figure. The numbers in the three circles at the bottom of the figure are 2, 7, and 8. Their sum is $2 + 7 + 8 = 17$.

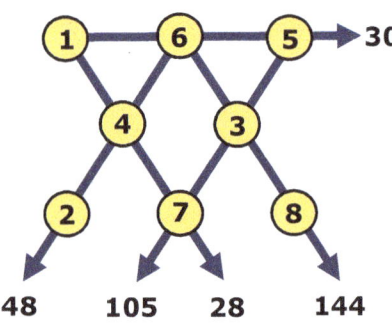

---

18 (B) 25%
Suppose the area of the union is 100 square units. Then the area of the intersection is 45 square units and the area of the triangle outside the circle is 40 square units. Hence, the area of the circle outside the triangle is $100 - 45 - 40 = 15$ square units. Finally, the percentage of the circle that lies outside the triangle is

$$\frac{15}{15 + 45} = 0.25 \times 100\% = 25\%$$

19 (D) 9
The shape consists of three non-overlapping hexagons: upper, left, lower. Each hexagon can be covered by 3 tiles in two ways. Just start with any triangle of the hexagon and move either clockwise or counterclockwise. This procedure gives us $2 \times 2 \times 2 = 8$ coverages of the shape.

However, it is also possible to tile the shape so that the tiles cross the boundaries of the hexagons. Starting with the pink tile and progressing clockwise across the whole shape, there is exactly 1 way to tile the rest of the shape. There is also exactly 1 tiling starting with the yellow tile, and 1 tiling starting with the turquoise tile. All three of these tilings are equivalent, meaning that there is only 1 tiling where at least 1 tile crosses a boundary of a hexagon.

Therefore, there are $8 + 1 = 9$ total tilings.

---

20 (B) $\frac{1}{5}$
Let the fraction of the route that Marc did by bike be b. Marc rides his bike for $20b$ minutes and he walks for $60(1 - b)$ minutes. Hence, $52 = 20b + (60 \times (1 - b))$, which gives $b = \frac{1}{5}$.

5 Point Solutions

21 **(B) 1**

The left upper square and the right upper square share entries in the middle column. So, the sum of entries of these 2 squares are equal only if the entry under 2 equals [(4 − 2) + the entry under 4]. Let x be the entry under 4. Then, $x + 2$ is the entry under 2. Apply the same argument to the left lower square and the right lower square. We see that the entry in the lower left corner of the table must be 1, since $(x + 2) + 1 = x + 3$. A complete table for the problem is shown, where x, y, and z are any numbers. The common sum is $x + y + z + 4$.

2		4
x+2	x	
		3

2	y	4
x+2	z	x
1	y+1	3

22 **(C) 80 km**

Village A is not between B and C, since $AC = 75$, which is > $BC = 20$. Village D is not between B and C, since $BD = 45$, which is > $BC = 20$. Fix the positions of B and C, so that C is to the right of B. Then, there are 4 options to locate A and D:

1. B is between A and C, so $AB + BC = AC$ or $AB + 20 = 75$. Hence, $AB = 55$. DB is given as 45, so A is further to the left than D. D is between A and B, so $AD + DB = AB$ or $AD + 45 = 55$. Hence, $AD = 10$. From left to right, the points are in order A, D, B, C and $AD = 10$, $DB = 45$, $BC = 20$.

2. C is between A and B, so $BC + CA = BA$ or $20 + 75 = BA$. Hence, $BA = 95$. BD is given as 45, so A is further to the right than D. D is between B and A, so $BD + DA = BA$ or $45 + DA = 95$. Thus, $DA = 50$. D is between C and A, so $CD + DA = CA$ or $CD + 50 = 75$. Hence, $CD = 25$. From left to right, the points are in order B, C, D, A and $BC = 20$, $CD = 25$, $DA = 50$.

3. A is furthest to the left and D is furthest to the right. From left to right, the points are in order A, B, C, D and it can be easily computed that $AB = 55$, $BC = 20$, $CD = 25$.

4. D is furthest to the left and A is furthest to the right. From left to right, the points are in order D, B, C, A and $DB = 45$, $BC = 20$, $CA = 75$.

In the diagram, the four cases are summarized graphically.

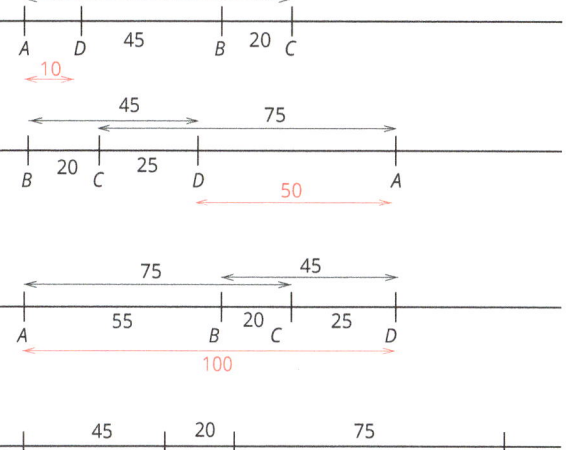

From the five given options, 80 km is the only option that cannot be the distance between A and D.

23 (D) $\frac{12}{7}$

Let h be the height and w the width of one small rectangle. Then, $AB = 4w$ and $BC = h + w$. So, $AB = DC = 3h$, or $3h = 4w$.

$$\frac{AB}{BC} = \frac{3}{3} \times \frac{4w}{h+w} = \frac{12w}{3h+3w} = \frac{12w}{4w+3w} = \frac{12}{7}.$$

---

24 (A) $\frac{5}{3}$ liters

The mixture has 1 liter extra of blue paint – namely, 3 liters of blue instead of 2. Intuitively, the best he could do is to throw away enough mixture to reduce its contents by 1 liter of blue, and fill up the rest with yellow. Since he wants to throw away $\frac{1}{3}$ of the blue paint, he must throw $\frac{1}{3}$ of the mixture away. Hence, he should throw away $\frac{1}{3} \times 5 = \frac{5}{3}$ liters of the green mixture. Formally, he can throw away m liters of the green mixture where $0 \le m \le 5$, add b liters of blue paint, and also add y liters of yellow paint, where $b, y \ge 0$. He wants to end up with 2 liters of blue paint and 3 liters of yellow paint. $(5 - m)$ liters of the green mixture contains $(5-m) \times \frac{2}{5}$ liters of yellow paint, and $(5-m) \times \frac{3}{5}$ liters of blue paint. Thus, the painter ends up with $[y + (5-m) \times \frac{2}{5}]$ liters of yellow paint, and $[b + (5-m) \times \frac{3}{5}]$ of blue paint. He wishes that $[y + (5-m) \times \frac{2}{5}] = 3$ and $[b + (5-m) \times \frac{3}{5}] = 2$, where $b, y \ge 0$ and $0 \le m \le 5$.
From the second equation $[b + 3 - \frac{3}{5} \times m] = 2$, so $\frac{3}{5} \times m = b + 1$, or $m = \frac{5}{3} \times (b+1)$. For $b \ge 0$, the smallest value of m is $\frac{5}{3} \times (0+1) = \frac{5}{3}$, and it is between 0 and 5. From the first equation, $[y + (5-m) \times \frac{2}{5}] = 3$, so $y + 2 - \frac{2}{5} \times m = 3$, or $y = 1 + \frac{2}{5} \times m$. Simplifying, $y = 1 + \frac{2}{5} \times \frac{5}{3} = \frac{5}{3}$.
$m = \frac{5}{3}$ makes sense, since we are not adding any blue paint ($b = 0$), and we want to have 5 liters of paint, so $\frac{5}{3}$ liters of yellow paint must compensate for $\frac{5}{3}$ liters of the green mixture. Therefore, the painter should throw away $\frac{5}{3}$ liters of the green mixture, and add $\frac{5}{3}$ liters of yellow paint.

---

25 (D) 54

Let the faces of the brick have areas A, B, and C, as shown. From the given surface areas of the shapes, we obtain $4A + 4B + 2C = 72$, $4A + 2B + 4C = 96$, and $2A + 4B + 4C = 102$. Adding, we get $10(A + B + C) = 270$. So the surface area of the original brick is $2(A + B + C) = 270 \div 5 = 54$.

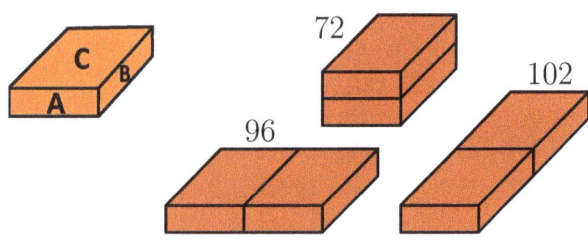

---

26 (B) 6

At least six colored squares are necessary so that each of the six 1×4 rectangles shown has at least one colored square, as we can see in the picture. Six colored squares are sufficient, as we can see in the picture — counting from each colored square to the top, right, bottom, and left, the maximum number of white squares is 3 — which means that there is no contiguous block of 4 white squares.

 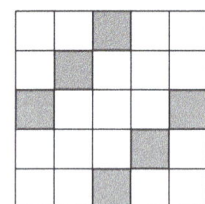

---

27 (A) Thursday

If today is one of the zebra's "truth-telling" days, then it must be Thursday, as this is the only truth-telling day preceded by a "lying" day. If today is one of the zebra's "lying" days, then the zebra's statement, "Yesterday was one of my lying days," is false, so yesterday was one of the zebra's "truth-telling" days. Today must be Monday, as this is the only lying day preceded by a truth-telling day. So from the zebra's response, Mowgli knows that today is either Thursday or Monday. By the same reasoning, from the panther's response, Mowgli knows that today is either Thursday or Sunday. Thus, for both the zebra's and panther's statements to be consistent, Mowgli must conclude that today is Thursday.

28 (C) 15

Let the initial number of points on a line be n. After Renard's first action, the number of points becomes $2n - 1$, as another $n - 1$ points are added, one between each adjacent pair of points. Renard repeats this action three more times: after the 2nd step, there are $2(2n - 1) - 1$ points; after the 3rd step, there are $2(2(2n - 1) - 1) - 1$; and after the 4th step, there are $2(2(2(2n - 1) - 1) - 1) - 1 = 16n - 15$ points. Therefore, $16n - 15 = 225$, and $n = 15$ points.

29 (E) 36

Let $\alpha = \angle BAD$. Triangle ADB is isosceles, so $\angle ABD = \alpha$. For triangle ADB, the external angle $\angle BDC = \angle BAD + \angle ABD$. So, $\angle BDC = 2\alpha$, or $\angle EDC = 2\alpha$. Let $\beta = \angle CBE$. Triangle BEC is isosceles, so $\angle BCE = \beta$. For triangle BEC, the external angle $\angle CED = \angle CBE + \angle BCE$. So, $\angle CED = 2\beta$. Triangle CED is isosceles, so $\angle CED = \angle EDC$, or $2\alpha = 2\beta$. Hence, $\alpha = \beta$. From triangle ABC, $\alpha + 2(\alpha + \beta) = 180°$. $\alpha = \beta$, so $5\alpha = 180°$, and therefore, $\alpha = 36°$. Notice that BD is the bisector of $\angle B$ and CE is the bisector of $\angle C$ in triangle ABC.

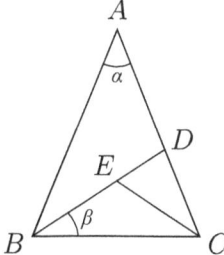

30 (B) 337

Suppose that there were initially no kangaroos, and you have to add them in one by one. For every koala in park 1, you add a kangaroo in parks 2, 3, 4, 5, 6, and 7. For every koala in park 2, you add a kangaroo in parks 1, 3, 4, 5, 6, and 7. For every koala in park 3, you add a kangaroo in parks 1, 2, 4, 5, 6, and 7, and so on. Notice that you have added 6 kangaroos for every koala in every park. Therefore, the number of kangaroos is 6 times the number of koalas. The number of koalas is 1/6 the number of kangaroos, which is $2022 \div 6 = 337$.

2024

2024

3 Point Solutions

1 (B)

Only for (B), two rings are formed that must pass through each other, and it is impossible to untie this knot without cutting the string.

---

2 (C)

Complete the lines and rotate the figure upside down. It can be seen that the center pentagon is

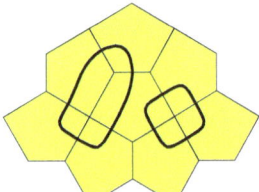 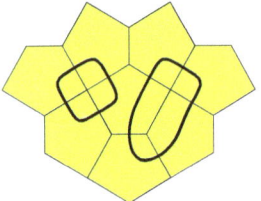

---

3 (E) 50%
The initial figure can be divided into 4 congruent right triangles, and the final figure into 6, as shown. Thus the percent increase in area is $(6 - 4) \div 4 \times 100\% = 50\%$.

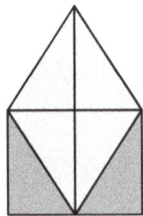

---

4 (D) 60

$$\frac{20 \times 24}{2 \times 0 + 2 \times 4} = \frac{20 \times 24}{0 + 8} = \frac{20 \times 24}{8} =$$
$$= 20 \times \frac{24}{8} = 20 \times 3 = 60$$

---

5 (D) 12
A regular tetrahedron has 4 vertices. Each cut removes one vertex but creates three new ones below it. Therefore, after four cuts we remove 4 vertices but create 4 × 3 = 12 new ones. Thus there are 4 − 4 + 12 = 12 vertices afterwards.

---

6 (B) 4
We can look at each possible option for the first counter:

- If the counter 5 is first, only the number 5111 can be formed.
- If the counter 1 is first, 1511 and 1115 can be formed.
- If the counter 11 is first, 1151 and 1115 can be formed.

Thus there are only 4 possible numbers: 5111, 1511, 1151, and 1115.

---

7 (E) Eva
Al must get .
Since everyone must get different fruits, Eva cannot get , so she must get 🍒.

---

8 (C) 5
With nine adults the elevator is at 9/12 = 3/4 of its weight limit, so the additional children can weigh at most 1/4 of the weight limit. Since 20 children take up the full weight limit, at most 20 × 1/4 = 5 children can be added to the elevator.

9. **(C) 13**
The numbers in the first column must be two different positive integers with a product of 4: the only such pair is (1, 4). If 4 is in the top-left cell, then the number in the top-right cell is such that 4 × ☐ = 6, which would mean this number is not an integer. Thus the top-left cell is 1, and the bottom-left cell is 4. From here we can fill out the rest of the grid as shown. The desired sum is 1 + 6 + 4 + 2 = 13.

1	6	6
4	2	8
4	12	

10. **(C) 78 cm**
To go from ten to four carts, we remove six carts and 168 − 108 = 60 cm. Thus removing one cart decreases the length by 10 cm. To go from four to one carts, we remove three carts and 30 cm, so the length of one cart is 108 − 30 = 78 cm.

4 Point Solutions

11. **(B) 4°**
Since each slice is 1/10 of the circle, the angle of each slice is 360° ÷ 10 = 36°. After removing one slice (with angle 36°), there are nine equal gaps. The combined angle of these nine gaps must equal the angle of the slice that was removed. Therefore, each gap has an angle of 36° ÷ 9 = 4°.

12. **(A)** | 1 | 1 | 3 |

One of the pieces has a complete row of four with sum 2 + 1 + 3 + 1 = 7, so each row and column must add up to 7. Therefore the sum of the completed square must be 7 × 4 = 28, since it has four rows, each adding up to 7. The three pieces given have totals of 7, 8, and 8, so the total of the missing piece must be 28 − 7 − 8 − 8 = 5. Of the options, only (A) has a total of 5. The completed square is shown.

2	1	3	1
2	2	2	1
1	3	1	2
2	1	1	3

13. **(A) 0 m²**
Consider two triangles: one covering both *A* and *C*, and one covering both *B* and *C*. They have the same area, since they share their base and their height. The difference in their areas is 0, but it is also (A + C) − (B + C) = A + C − B − C = A − B. Thus A − B = 0 m².

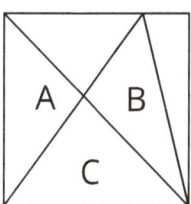

14. **(D) 52**
The only way a chick could have eaten 44 fish in total is to have eaten 5 fish six times and 7 fish twice. Therefore, it had eaten fish on 8 days. The total number of fish Paula brought back on those 8 days was 8 × 12 = 96. Hence, the number of fish eaten by the second chick was 96 − 44 = 52.

15. **(A) 18**
There are 30 faces to cover. Adding a cube into one of the "angles" about one edge of the innermost cube of the structure covers 2 faces at the same time. There are 12 edges of this innermost cube, hence with 12 cubes you cover these 24 inner faces. To cover the outermost faces, you need 6 more cubes, hence 6 + 12 = 18 cubes are needed altogether.

16. **(D) 3036**
Let *u* denote the number of uphill jumps and *d* the number of downhill jumps. Since the kangaroo covers three times the distance with each downhill jump as it does with each uphill jump, it will make three times as many uphill jumps as it does downhill jumps. Therefore $u = 3d$. Since we are told that $u + d = 2024$, we have $3d + d = 2024$ or $4d = 2024$, so $d = 2024 ÷ 4 = 506$. Therefore the total distance, in meters, that the kangaroo jumps is 506 × 3 + 3 × 506 × 1 = 3036.

17 (B) 10

Consider the rectangles with perimeters 16 and 18. Together, their perimeter is $(2a + 2d) + (2b + 2c)$. Consider the rectangles with peimeters 24 and ? . Together, their perimeter is $(2b + 2d) + (2a + 2c)$.

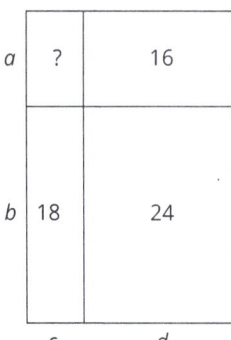

These quantities are the same (both equal to $2a + 2b + 2c + 2d$, the perimeter of the large rectangle). Thus $16 + 18 = 24 + ?$, so $? = 10$.

---

18 (C) 75

Consider 100 grams of fresh mushrooms. Since 80% of this is water, the remaining dry matter has a mass of 20 grams. When dried, this 20 grams of dry matter now represents 80% of the total mass, and so the 20% that is water has a mass of 5 grams (since 20% is one-fourth of 80%). Therefore the total mass of the dried mushrooms is $20 + 5 = 25$ grams, and hence the decrease is 75 grams or 75% of the original mass.

---

19 (D) 6000

Each hexagon borders six triangles. However, we cannot just multiply the number of hexagons by six: since each triangle borders three hexagons, each triangle will be counted three times. Therefore, 3 × the number of triangles = 3000 × 6, so the number of triangles is $(3000 × 6) ÷ 3 = 6000$.

20 (E) 9

List out all the possibilities for each person:

Alexa: (1, 5), (2, 4)

Bart: (1, 6), (2, 7), (3, 8), (4, 9)

Clara: (2, 9), (3, 6)

Diana: (1, 2), (2, 4), (3, 6), (4, 8)

If Alexa picks (2, 4), Clara must pick (3, 6), but then all of Diana's options are taken. Therfore, Alexa must pick (1, 5). The remaining options are:

Alexa: (1, 5)

Bart: (2, 7), (3, 8), (4, 9)

Clara: (2, 9), (3, 6)

Diana: (2, 4), (3, 6), (4, 8)

If Clara picks (2, 9), then Bart must pick (3, 8), but then all of Diana's options are taken. Thus, Clara must pick (3, 6), so Diana must take a 4. Then Bart can only take (2, 7), and Diana must take (4, 8). The unused number is 9.

5 Point Solutions

21 **(A) 9**

Let's count the number of vertical and horizontal segments of each digit:

Digit	0	1	2	3	4	5	6	7	8	9
Horizontal	2	0	3	3	1	3	3	1	3	3
Vertical	4	2	2	2	3	2	3	2	4	3

For there to be 5 horizontal segments, we need to consider the ways three numbers from 0 to 3 can add to 5: 2 + 2 + 1, 3 + 1 + 1, 3 + 2 + 0. The first way is impossible since only one digit has 2 horizontal segments and we need all digits to be distinct. The second way is not also possible, since only 4 and 7 have one horizontal segment, and the digits 4 and 7 have a total of 3 + 2 = 5 vertical segments, meaning the third digit would need to have 5 vertical segments and the maximum possible is 4.

Therefore there must be 3, 2, and 0 horizontal segments in the digits. Only digit 1 has no horizontal segments and only digit 0 has two horizontal segments, and they have 4 + 2 = 6 vertical segments between them. Therefore the final digit has 3 horizontal segments and 10 − 6 = 4 vertical segments, so it must be the digit 8. The requested sum is 1 + 0 + 8 = 9.

22 **(E) 6**

A square has four lines of symmetry. Reflect the shaded cells about each line of symmetry, as shown:

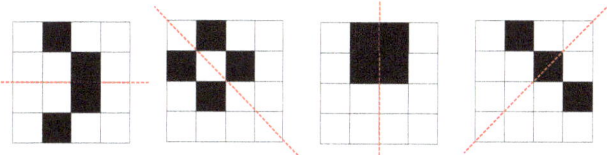

In the first three cases, there are already four shaded cells, so there is only one way in each of those cases. In the last case, we need to shade one additional cell to keep the line of symmetry, so we must shade one of the cells on the line of symmetry (the red dashed line). There are three unshaded cells on this line, so there are three ways to complete the diagram in this case. The final count is 1 + 1 + 1 + 3 = 6.

23 **(B) 92**

Let h be the height of the rectangle. Then the radii of the three semicircles are h, $h − 5$, and $h − 7$. Therefore, by considering the length of the rectangle, the diameters of the semicircles add up to 36, so $2(h + h − 5 + h − 7) = 36$. Expanding, we get $6h − 24 = 36$, or $h = 10$. Therefore the perimeter of the rectangle is $2 × (36 + 10) = 92$.

24 **(D) 25**

In order for a student to receive the ball again after passing it on, the ball must bypass a multiple of 50 students. Since the ball is caught by every 6th student, it also bypasses a multiple of 6 students at any given time. Therefore, the ball must bypass 150 = LCM(50, 6) students for the first student to receive the ball again.

Before this occurs, the ball has been passed to 150 ÷ 6 = 25 students (since it bypasses 6 students with each step), and we can be sure that no two students are the same because this would take at least 150 passes.

After the first student receives the ball again and starts passing it around for the second time, the passing pattern repeats itself, and the same 25 players will get the ball again, meaning the other 25 players will never touch the ball.

25 (D) 6

Let a and b be the missing values in the boxes in the left-hand side and right-hand side of the bottom row respectively. Then the values in the boxes in the middle row are an and bn and the value in the top row is abn^2. Therefore, $abn^2 = 720$; in other words, n^2 is a perfect square divisor of 720. So the answer (the number of possible values of n) equals the number of perfect squares that divide 720.

The largest perfect square that divides 720 is 144, since $720 \div 144 = 5$. Therefore we need to check all perfect squares at most $12^2 = 144$ and see if they divide 720. Looking through them, only the perfect squares $1^2, 2^2, 3^2, 4^2, 6^2, 12^2$ divide 720, so the answer is 6.

26 (E) 29

After the customer buys one basket, let x be the number of duck eggs. Then there are $2x$ chicken eggs, so in total there are $x + 2x = 3x$ eggs; that is, the number of eggs is a multiple of 3.

Initially, Frank had a total of $4 + 6 + 12 + 13 + 22 + 29 = 86$ eggs. Since 86 leaves a remainder of 2 when divided by 3, the number of eggs sold also has a remainder of 2 when divided by 3, because their difference leaves a remainder of 0 (that is, it is a multiple of 3). Of the options, only 29 leaves a remainder of 2 when divided by 3.

27 (D) 90°

Using parallel lines, we can construct a triangle with angles $α$, $β + γ$, 90°, as shown in the diagram. Hence $α + β + γ = 90°$.

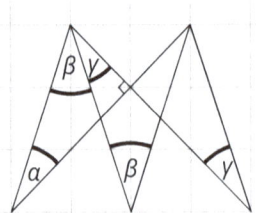

28 (B) Al

Assume the people are correct one at a time, and look for contradictions:

- Suppose Tom told the truth. Since the total number of coins is 30, his response for gold coins would be $30 - 9 - 11 = 10$. However, this is the same as Pit's response for gold coins. Since the other three pirates were lying in all three of their answers, this is not possible.

- Suppose Pit told the truth. His response for silver coins would be $30 - 10 - 10 = 10$, but this is the same as Jim's response, so this is not possible.

- Suppose Jim told the truth. His response for bronze coins would be $30 - 9 - 10 = 11$, but this is the same as Tom's response, so this is not possible.

Therefore, since none of Tom, Pit, or Jim told the truth, only Al can have told the truth. His missing response for silver coins would be $30 - 7 - 12 = 11$, and it can be seen that none of his answers match any of the other pirates' answers.

29 (C) 30 min

Say Alex starts at *A*, Bob starts at *B*, the first meeting is at *M*, and the second meeting is at *N*. Alex's path is shown in red above the line, Bob's path is shown in blue.

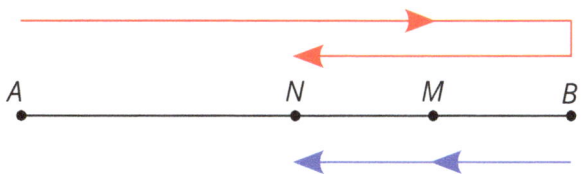

Let V_A and V_B be Alex's speed and Bob's speed respectively. Since $V_A = 3V_B$, then $AM = 3BM$. Each person covered this distance in 15 min. Since $V_A \div V_B > 2$, the second meeting took place before Bob reached *A*. Between meetings, Alex covered the distance $MB + BM + MN$, while Bob covered the distance MN (the second arrow in each path above). Since $MB + BM + MN = 3MN$, we have $MB = MN$. Thus, Bob covered the distance $BM + MN = 2BM$ from the start till the second meeting in $2 \times 15 = 30$ min.

30 (A) 45

Since $\angle A = \angle B = 90°$ and $AE = BC$, $ABCE$ is a rectangle. Draw the lines *FG*, *HI*, and *JK* parallel to *AB*. The figure is symmetric about *PD*, where *P* is the midpoint of *AB*.

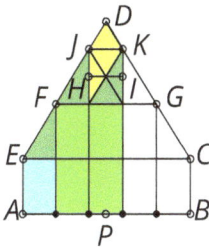

Let's look at the areas of the given figures: comparing the light and dark gray areas, we get:

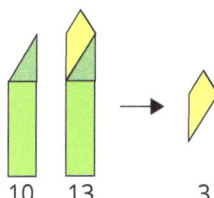

The leftover piece can be divided into six congruent right triangles of area 0.5.

From this we can get that the rectangles between *CE* and *FG* have area 4:

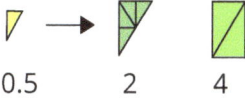

Looking at the leftmost trapezoid, we get that each of the five rectangles that *ABCE* is cut into has area 4:

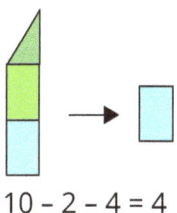

$10 - 2 - 4 = 4$

Therefore the area of the leftmost trapezoid is $4 + 2 = 6$. By symmetry, the area of the rightmost trapezoid is also 6, so the total area of the pentagon is (adding the five regions given in the problem one at a time) $6 + 10 + 13 + 10 + 6 = 45$.

Part III
Answer Key

	2006	2008	2010	2012	2014
1	B	B	C	B	D
2	D	C	C	D	D
3	D	B	D	A	A
4	E	C	A	E	B
5	D	C	B	C	E
6	B	E	C	E	E
7	A	B	E	C	B
8	D	B	B	D	E
9	D	E	B	D	B
10	E	A	C	D	D
11	B	C	C	C	E
12	A	B	D	C	B
13	C	A	D	B	C
14	D	C	B	C	E
15	B	A	C	C	E
16	B	A	A	D	C
17	B	C	A	C	B
18	C	A	B	E	E
19	A	D	E	D	D
20	B	D	B	B	A
21	E	D	C	E	E
22	B	C	C	D	D
23	E	C	E	B	B
24	D	D	B	D	D
25	B	C	C	D	C
26	C	A	C	E	A
27	D	B	B	B	B
28	E	D	A	C	C
29	C	C	D	C	A
30	E	B	D	A	B

	2016	2018	2020	2022	2024
1	C	B	B	E	B
2	A	E	A	B	C
3	C	B	C	C	E
4	D	D	A	B	D
5	B	C	E	C	D
6	A	D	E	A	B
7	C	C	D	D	E
8	C	C	B	B	C
9	B	D	D	A	C
10	C	D	B	C	C
11	E	B	A	D	B
12	E	C	C	D	A
13	C	D	B	E	A
14	B	C	C	A	D
15	B	A	C	C	A
16	D	D	C	B	D
17	D	D	A	D	B
18	B	B	D	B	C
19	D	A	B	D	D
20	B	B	E	B	E
21	D	E	A	B	A
22	D	C	A	C	E
23	E	B	D	D	B
24	D	B	B	A	D
25	E	D	C	D	D
26	C	C	C	B	E
27	A	B	E	A	D
28	E	E	D	C	B
29	C	C	D	E	C
30	A	C	C	B	A